T0140565

Rise of the Self-Replicators

Tim Taylor • Alan Dorin

Rise of the Self-Replicators

Early Visions of Machines, AI and Robots
That Can Reproduce and Evolve

 Springer

Tim Taylor
Independent Researcher
Edinburgh, UK

Alan Dorin
Faculty of Information Technology
Monash University
Clayton, VIC, Australia

ISBN 978-3-030-48233-6   ISBN 978-3-030-48234-3   (eBook)
https://doi.org/10.1007/978-3-030-48234-3

This Springer imprint is published by the registered company Springer Nature Switzerland AG
The registered company address is: Gewerbestrasse 11, 6330 Cham, Switzerland

Preface

The origins of this book date back to 2014, during a period when one of the authors (TT) was working in the other's (AD's) lab at Monash University in Australia. We had first met at a conference on Artificial Life in 2002; our shared interests in the subject meant that we had kept in touch regularly since then, despite living on opposite sides of the world (TT in Edinburgh, Scotland, and AD in Melbourne, Australia).

Artificial Life (or "ALife" for short) is the application of biological design principles for building complex, intelligent systems. It might be studied in software, hardware or "wetware" (molecular systems), and might be used for a variety of purposes. The two most common reasons people pursue ALife are as an approach to understanding biological systems and as an approach to building intelligent robots and artificial intelligence (AI) systems. It is also used by philosophers, social scientists, artists, and many others besides.

The ALife conferences are an exciting interdisciplinary melting-pot of ideas. At the conference in 2002, we soon discovered that we shared very similar interests in designing artificial worlds in software. We both used computational analogues to the processes of biological self-reproduction, evolution and natural selection to populate our worlds with interesting creatures. We had both completed PhDs in this area in the 1990s. Beyond our experimental work, we also shared an interest in early mechanical models of living systems, and in the history of thought about technology inspired by biology.

During our period of working together in 2014, our original inspiration for writing this review came from reading a recently published account [40] of Alfred Marshall's ideas of machine self-reproduction and evolution from the 1860s (we discuss his work in Sect. 3.2). Although we were aware of Samuel Butler's writing on the subject at around the same time (Sect. 3.1), we had not come across Marshall in this context before. We were therefore curious whether there might be other work from around this time, or even earlier, that discussed such ideas.

Another motivation was to highlight the pioneering work in the early 1950s of Nils Aall Barricelli on self-reproduction and evolution in software (Sect. 5.2.1). While John von Neumann's theoretical work on self-reproducing machines from around that time is widely discussed in the literature (Sect. 5.1.1), personal experience suggested that Barricelli was still a relatively unknown figure within the ALife community, despite having a strong claim to be regarded as one of the field's founding fathers.

Our purpose in writing this book was therefore to review the early history of the idea of self-reproducing and evolving machines, tracing it back as far as we could. This being the case, much of the book (Chaps. 2–6) is written as a guide to the literature on the subject, presented in chronological order from the earliest inklings of the idea up to the present day. While we provide some commentary and suggest classifications of the work in terms of the goals of the authors we survey, our primary aim is to present a comprehensive archive of thought about self-reproducing machines. These chapters represent the most extensive early history of the subject published to date and include coverage of many works that have not been widely discussed elsewhere. We do provide a synthesis and summary of the concepts discussed in Chap. 7, and it is there that we offer more of our own views on the field and where we see it heading.

The audience we have in mind includes anyone wishing to learn about the origins of the idea of self-reproducing and evolving machines, especially those interested in drawing lessons from this early work regarding likely future developments in the field. Most obviously, the audience will be Artificial Life and Artificial Intelligence practitioners. We also believe the subject will be of interest to many philosophers, biologists, engineers, historians of science, and those working in the emerging field of AI safety and ethics.

We hope the content will be of value in informing a wider general readership too. For that reason, in Chap. 1 we discuss the profound future implications of the technology and explain why it is a subject of broad relevance. We have tried to make the text accessible and to avoid technical jargon, although this has not always been possible. In particular, in Chap. 5 we discuss at greater length the details of the first realisations of self-reproducing machines in the 1950s, and in Chap. 7 we summarise technical aspects of the design of self-reproducing machines. Nevertheless, we hope we have found a reasonable balance between technical detail and accessibility, even in these sections.

Having conceived the idea of the book in 2014, the content was primarily researched and written by TT, in between other work, over the next five years. AD provided feedback and ideas during numerous discussions over that period, together with detailed comments and editorial suggestions on drafts of the book.

At least at the time of writing, the possibilities of self-reproducing and evolving machines are not commonly addressed in popular discussions about robotics and artificial intelligence. However, as you will see in what follows, we argue that work in this area has potentially huge implications for the future of humanity. We hope this book plays a small part in bringing these intriguing and important topics back onto the agenda when considering the deep history and far future of intelligent machines and the fate of our own species.

Additional information and materials relating to the book can be found online at
http://www.tim-taylor.com/selfrepbook/.

Edinburgh & Melbourne,   *Tim Taylor*
April 2020   *Alan Dorin*

Acknowledgements

The task of researching and preparing this book has been considerably aided by generous assistance from a variety of sources.

Our efforts in trying to trace an original authoritative source for the story involving René Descartes and Queen Christina of Sweden (Sect. 2.1) were greatly helped by invaluable feedback and suggestions from many experts in the field: John Cottingham, Stephen Gaukroger, Minsoo Kang, Jessica Riskin, Gary Hatfield, Justin E. H. Smith, Roger Ariew, Theo Verbeek, Erik-Jan Bos, Susanna Åkerman, Mike Wheeler, Robert Freitas, Martyn Amos and Moshe Sipper. We are extremely grateful to all of them.

Nancy Henry was very helpful in providing information and feedback as we explored the work of George Eliot, and we very much appreciate her assistance. Likewise, we are very grateful to Nora Eibisch for her guidance as we explored the work of Konrad Zuse.

We also thank Justin E. H. Smith for bringing Bernard Le Bovier de Fontenelle's work to our attention (Sect. 2.1), Seth Bullock for highlighting Alfred Marshall's contribution (Sect. 3.2), and Jeremy Pitt for pointing us to Robert Sheckley's sci-fi work (Sect. 4.1.3).

Our research activities benefited greatly from the facilities and helpful staff at the National Library of Scotland, the Special Collections staff at the University of Edinburgh library, Monash University Library, the online resources of the National Library of New Zealand, the online resources of the Wellcome Library in London, Kathryn Mouncey and Jane McGuinness at the British Library, Rosemary Clarkson at the Darwin Correspondence Project of the University of Cambridge and Dagmar Spies at the University Archives of the Technical University of Berlin.

Some small sections of Chaps. 3, 4 and 7 were previously published in a conference paper presented at the 2018 Conference on Artificial Life [282]. That publication is copyright MIT Press, and the text is reproduced here in accordance with the publication agreement between MIT Press and the authors.

We gratefully acknowledge the use of Overleaf (https://www.overleaf.com/), the free online collaborative LATEX authoring tool, in preparing the draft of this work.

Image Credits

Contents

Chapter 1
Self-Reproducing Machines: The Evolution of an Idea

> Within the next century we will likely witness the introduction on earth of living organisms originally designed in large part by humans, but with the capability to reproduce and evolve just as natural organisms do. This promises to be a singular and profound historical event—probably the most significant since the emergence of human beings.
>
> J. Doyne Farmer and Alletta d'A. Belin, *Artificial Life: The Coming Evolution*, 1991 [109, p. 816]

In the chapters that follow, we explore the early history of thought about machines that can reproduce and evolve. Unless you work in one of a small set of rather specialised academic or engineering disciplines, you may not have come across much discussion of these ideas before, at least not beyond the realm of sci-fi films and novels. We believe that will soon change as the work whose origin we describe here progresses.

1.1 Two Central Questions

There are two underlying questions that have provided the central motivation for all of the work that we cover:

- Is it possible to design robots and other machines that can reproduce and evolve just like biological organisms do?
- And, if so, what are the implications: for the machines, for ourselves, for our environment, and for the future of life on Earth and elsewhere?

Out of these two questions spring many others. Might this be a route by which we could create machines whose capabilities go beyond the rather narrow focus of today's artificial intelligence (AI) systems, and which automatically evolve towards a more powerful and wide-ranging artificial general intelligence (AGI)? In contrast

© Springer Nature Switzerland AG 2020
T. Taylor, A. Dorin, *Rise of the Self-Replicators*,
https://doi.org/10.1007/978-3-030-48234-3_1

to today's AI systems, might evolution and natural selection instil inner desires and purpose in these machines, as it has done in the biological realm? More futuristically, could a spaceship that can build copies of itself from raw materials scavenged from asteroids and other planets be a route by which we could travel immense distances to explore and colonise worlds in other solar systems or even other galaxies? And what of the economic disruption that might be caused here on Earth by a nanoscale manufacturing plant that could autonomously build more copies not just of its output produce but of the manufacturing plant itself, all at an exponentially increasing rate?

From even a cursory consideration of these questions, it is clear that technologies like these could potentially pose serious threats to our environment, and even to the future of the human race itself. And yet, perhaps instilling machines with the power of reproduction and evolution might be the most promising approach to ensure the survival of intelligent life in the far future of the Earth and elsewhere in the universe?

1.2 The Promise of Self-Reproducing Machines

Oh my goodness, shut me down! Machines making machines—how perverse!
                                        C-3PO, *Star Wars: Episode II Attack of the Clones*, 2002 [193]

We realized that the true problem, the true difficulty, and where the greatest potential is—is building the machine that makes the machine.
                                        Elon Musk, *Tesla Annual Shareholder Meeting*, 2016 [213]

I knew that baby meant we are more than just slaves. If a baby can come from one of us, we are our own masters.
                                        Freysa (a Nexus-8 replicant), *Blade Runner 2049*, 2017 [108]

At the heart of all of these questions is the concept of a *self-reproducing machine* (or *self-replicator* for short). Is it conceivable that we might be able to design and build such machines in the near future? Take a moment to look at any modern gadget you own: your mobile phone, your television, your bicycle, your toaster. Think about all the components from which it is made. The chances are, every single component has been manufactured by another machine. The idea of *machines making machines* is so commonplace that we rarely stop to give it a second thought. Likewise, the raw materials required for each component have been mined and processed by other machines. The transportation of raw materials to factories, and of manufactured goods to warehouses and stores, is achieved by yet more machines. At every step, the role of humans is becoming increasingly redundant. Recently developed technologies ranging from 3D printing to self-driving vehicles are enabling every aspect of the manufacturing process to become more and more automated.

Could we someday reach the point where the *entire* manufacturing process is completely automated, from the mining of raw materials to the delivery of a new gizmo to your front door? What are the limits of what could be manufactured by completely autonomous machines? Given where we are at present, it doesn't seem too far a leap to imagine a machine that could manufacture a complete copy of itself, if provided with the necessary parts. If you find this hard to imagine, how about a group of machines: could we design a large number of different machines that, between them, collectively build a copy of every machine in the group?

If such a feat were achievable, how about finessing the design so that the machine also collects and processes the raw materials required to build its offspring? Although there are doubtless major engineering challenges to be overcome to realise this, there are no obvious reasons why this could not be accomplished in theory. An autonomous system like this would be a completely self-reproducing machine. Just like a biological organism, in appropriate conditions it would be able to produce offspring of its own kind.

The questions posed at the start of the chapter allude to some of the many profound applications of self-replicator technology. What might be the ultimate outcome of all of this, for the machines, for the environment, and for us?

1.3 Diverging Visions in the Early History of the Idea

These kinds of questions, and, indeed, the very idea of a self-reproducing machine, might seem like very modern conceptions. In fact, they have captured the imagination of scientists, philosophers, writers and the general public for hundreds of years. One of our primary purposes in writing this book has been to explore the very early history of these ideas.

Our search has taken us back as far as the late 1600s, as the full implications of René Descartes' views of animals as machines began to be explored and debated. For the first couple of centuries that followed, most of the discussion was centred on the question of whether it was possible to design a machine that could reproduce by building a copy of itself. Throughout this book we refer to machines that possess this basic capacity for building a faithful copy of themselves as *standard self-replicators*, or **standard-replicators** for short.

In the mid-nineteenth century, a pivotal development in the history of the subject was triggered by the publication of Darwin's *On the Origin of Species* in 1859 [72]. Within a decade of its appearance, we find multiple extended discussions of the possibility of machines that can not only reproduce, but can also evolve by the natural selection of heritable variations to become better adapted, smarter and more complex over time. We refer to these kinds of machines as *evolvable self-replicators*, or **evo-replicators** for short. The early 1900s saw increasing speculation on these ideas, both by scientists and by sci-fi authors, and the 1950s saw the first implementations of simple self-reproducing systems in hardware and software.

The Darwinian vision of a machine that could reproduce and evolve like a biological species continued to inform a significant strand of work on self-reproducing machines as we move from the 1950s to more recent developments. This work embraced the possibility of heritable changes or mutations occurring in a self-replicator's offspring, these being the source of variety upon which Darwinian natural selection acts. Practical applications of evo-replicator technology include its use as a potential route for the automatic creation of advanced AI systems of far greater power than could be designed by humans, and also as an experimental tool by which we might better understand the conditions that have led to the evolution of intelligent biological life on Earth.

At the same time, as people began to think more seriously about the practical realisation of self-reproducing machines in the 1940s and 1950s, we also begin to see the emergence of a third distinct direction of work. This line of research was based on the insight that, given the right design, a self-replicator could quite easily be directed to produce specific goods and materials for us, in addition to reproducing itself. Thus, this work focused much more on the potential of physical self-reproducing machines as general-purpose manufacturing systems that could be deployed cost-effectively in inaccessible locations on Earth or further afield. The machines could then be remotely directed to produce a wide variety of specific outputs—think of them as glorified 3D printers—without requiring human maintenance, and with the potential to further replicate their activities elsewhere. We will refer to these kinds of systems as *manufacturing self-replicators*, or **maker-replicators** for short.

The key advantage of a maker-replicator's capacity for self-reproduction is that its creators would (in theory) only have to produce one of them initially, which could then automatically ramp up its activities *in situ* by producing more copies of itself in its target location. Thus, initial manufacturing and deployment costs are reduced by virtue of requiring only a single machine to commence the process. Furthermore, ensuring the long-term reliability of a single machine in a remote location becomes less of a crucial issue if the machine is able to make further copies of itself before it suffers a failure.

Another feature of the technology, which may be regarded as a benefit or a curse depending upon one's perspective and goals, is that it has the potential—given sufficient raw materials as input—to produce output at an ever accelerating rate. As a self-replicator builds more copies of itself, and the copies build more copies of themselves, the total number of machines could increase from the initial one, to two, then four, eight, sixteen..., the population doubling each time. From a human perspective, having paid the one-off cost of producing the first machine, we would have a process where the rate of output would increase exponentially over time! Everything we know about economics would be turned on its head.

Practical uses of this technology might include the economic large-scale production of valuable resources, for example converting sea water into fresh water, or building devices to capture solar energy. It might also be applied to geoengineering projects to tackle global warming. As we'll see in later chapters, some authors have proposed using maker-replicators to mine and process valuable resources on asteroids subsequently to be transported back to Earth, or even to terraform other planets prior to colonisation by humans.

Key issues in maker-replicator development include the engineering, economic and safety challenges in designing this kind of system. In contrast to evo-replicator development, those working on maker-replicators usually view the potential of evolution in their machines—and hence the possibility that the human designers might lose control of them as the machines develop unexpected abilities—as a danger that should be avoided through the design of appropriate safeguards. Having said that, as we'll see in Sect. 5.1.1, the development of the first significant theoretical work on self-replicators, by John von Neumann, addressed the design of machines that could *both* manufacture a wide variety of products *and* evolve to become more complex over time—that is, von Neumann's work was about evolvable manufacturing self-replicators (**evo-maker-replicators**).

There has also been a more limited amount of research on software maker-replicators. This work seeks to produce software self-replicators that have a general capacity to perform other specified computational tasks as well as reproduction. As with their hardware counterparts, a motivation for developing software maker-replicators is to create systems that can perform with high reliability in environments where they cannot be easily maintained by human operators. At first glance it might seem that safety issues associated with software maker-replicators are less critical than those of their hardware counterparts. However, as we discuss in Chap. 7, we should not underestimate their potential to cause serious harm not just online but also in the real world.

Following the first implementations in the 1950s, research in recent decades has seen many developments in all of these different flavours of self-replicator technology. The focus of work has shifted from speculation and science fiction to detailed studies of the design and implementation of self-reproducing machines in hardware and in software. During this time, work in designing physical self-reproducing machines in hardware has been largely concerned with maker-replicator systems, while software implementations mostly focus on evo-replicator systems. In both hardware and software we also see a significant amount of work on standard-replicators as a foundation upon which to progress towards the other two kinds of replicator. As we'll see later on, work on hardware maker-replicator systems is mostly still at the conceptual and prototype stage. On the other hand, work on software evo-replicator systems is more advanced, with an active and growing group of researchers working on improving the evolutionary potential of their implementations.

1.4 Relevance Today

In recent years, the idea of self-reproducing and evolving machines has been over-shadowed in the media by impressive breakthroughs in other areas of AI and machine learning with more immediate practical relevance. However, the long-term implications of self-replicator technology are potentially far more transformational.

We have already alluded to some possible applications of self-reproducing machines in the preceding pages. The introduction of real-world maker-replicator technology has the potential to revolutionise the production of materials on Earth, to provide an economical means of mining the resources of other moons and planets, and even to act as a route by which humankind—or our technological offspring—may explore and colonise worlds beyond our own solar system. In short, this technology could profoundly reshape our society, our relationship with the environment, and our place in the universe.

While most of the early work we describe in the following chapters considered large-scale (i.e. approximately human-scale or larger) versions of this technology, more recent developments in molecular-level systems such as nanobots provide an alternative medium in which physical maker-replicators could be instantiated. It is likely that research on systems at this scale will produce significant results before work at larger scales. At the same time, guarding against the risk of a runaway exponential self-replication process is more challenging at the smaller scale. If not carefully managed, physical self-replicators at both small and large scales bring with them substantial risks of causing catastrophic damage to the environment and existential threats to biological species, including our own.

As we show in the chapters that follow, it appears that the barriers to building physical maker-replicator systems are chiefly technical and economic rather than theoretical. The truly transformative potential of self-reproducing technology for commercial and sociological goals—and the potential financial returns that could be captured by a first-mover in the field—mean that we must assume that, sooner or later, some research group or corporate organisation will be successful in manu- facturing an operational physical self-reproducing system. While this is unlikely to happen in the near-term (especially for larger-scale systems), we expect significant progress in this area over a time frame of several decades. It is unclear whether, in the long run, this will be a positive or negative development for humanity. One thing that is clear, however, is that only by thinking carefully about these machines and their associated risks and implications will we be able to guide their future, and ours.

The development of software evo-replicator technology is the area of most rel- evance in the short-term, not least because systems already exist that implement evolutionary self-replicators in software. This technology holds the promise of sur- passing current commercial AI, by evolving systems with capabilities that go far beyond what they were originally designed to do. Some see evo-replicators as the most likely route by which to achieve human-level, or even superhuman-level, arti- ficial general intelligence (AGI). Research on instilling software evo-replicator sys- tems with the ongoing creative power of biological evolution is currently a central focus of research within the academic field of Artificial Life (ALife); this quest for *open-ended evolution* has recently been identified as a "grand challenge" for the field [269]. Although software-based evo-replicators don't present the same level of danger of environmental damage as posed by physical maker-replicators, they nevertheless have the potential to cause havoc in the online world if not properly managed. Recent trends in computer viruses that are designed specifically to cause damage to real-world infrastructure also demonstrate the dangers of underestimat- ing the potential negative effects of software-based evo-replicators.

Given the potential significance of all of the lines of research outlined above, it is vital that these developments are accompanied by a proper consideration of the possible risks and benefits involved. An appropriate starting point for this endeav- our is a careful consideration of what has been achieved to date, and what are the motivations and goals of those involved. Our review of the early history of the subject, as set out in the following chapters, is our contribution to constructing a firm foundation for these considerations.

1.5 A Note on Scope and Terminology

Before proceeding, some clarification is required on the scope of our review, on how this book differs from other reviews, and on our use of terminology.

As we describe in Chap. 5, the towering figure in the theory of self-reproducing machines is the Hungarian-American polymath John von Neumann, whose work on the topic in the late 1940s and early 1950s put the subject on a firm theoretical footing. Existing reviews of the subject generally start with von Neumann and concentrate on developments from the 1960s onward. In contrast, our focus in this book is on the earlier and less well-known history of the subject. After discussing the earlier work at length, we also summarise more recent developments and provide an introduction to more detailed reviews. By the end of the book, we therefore aim to have provided a sound overview of the entire history of the field, with pointers to further information about the more recent work.[1]

This is *not* a review of mechanical models of living things in general (a topic which has over two thousand years of history [60]), nor does it cover the much wider and more general idea of the evolution of technology.[2] Our focus is specifically on self-reproducing and evolving machines—we cover both physical machines (e.g. clockwork automata, electromechanical robots, and molecular-scale devices)[3] and logical machines (e.g. software programs and abstract automata), but we do not branch deeply into the area of bio-mechanical hybrids, bionics and cyborg technology where the reproductive functions remain predominantly biological.

Regarding our use of terminology, we acknowledge from the outset that the term *self-reproduction* can be problematic. No system is truly *self*-reproducing: the process is always the result of an *interaction* between a suitable structure and a suitable

---

[1] We acknowledge that our literature search has been conducted primarily in the English language. While we have spent some time searching for sources in other languages (including French, German, Spanish and Russian), we cannot rule out the possibility of the existence of relevant non-English language work in addition to those that we have found and cover here.

[2] Excellent coverage of these broader topics can be found elsewhere: for a review of mechanical models of living things, see, e.g., [250, 201], and for discussion of the general idea of the evolution of technology, see, e.g., [21, 6, 205].

[3] We use the terms *automaton* and *robot* more or less synonymously throughout the book. Both are machines driven by their own internal instructions and source of movement. The term *automaton* (plural *automata*) carries with it more of a sense that the device is acting in a rote fashion according to predefined rules; historically, it is the term that has been applied to clockwork models of living beings, and also to the kind of computational *cellular automata* models employed by John von Neumann (Sect. 5.1.1) and Nils Aall Barricelli (Sect. 5.2.1). *Robot* is a more recent term (see Sect. 4.1.2) and usually implies a autonomous machine with more intelligence than a simple automaton. While *automaton* may refer to a machine implemented either in hardware or in software, we reserve the term *robot* strictly to refer to hardware devices.

environment, causing the production of further copies of the structure. Many of the authors discussed in the following sections have offered insights into the various issues involved. We highlight these as we proceed and provide further discussion of the topic in Chap. 7.

With that said, we use the term *self-reproducing machine* to refer to a machine (or manufacturing plant) that, within a defined range of environments, can manufacture a copy of itself after collecting and processing the required raw materials. The term "self-reproducing machine" can be something of a mouthful when used frequently, so we also use the term *self-replicator* as a slightly shorter synonym.[4]

As stated above, we use the more specific terms *standard-replicators*, *evo-replicators* and *maker-replicators* to refer to particular flavours of work on self-reproducing machines. The focus of work on maker-replicators (manufacturing self-replicators) is on their ability to manufacture a wide range of products in addition to being able to produce copies of themselves. Those working with maker-replicators are generally very concerned that these machines work in a very controlled manner and are not able to evolve new capabilities. In contrast, the focus of work on evo-replicators (evolvable self-replicators) is very much on their ability to evolve and to acquire capabilities beyond those originally given to them by their human designers.

1.6 Outline of the Rest of the Book

In the chapters that follow we discern three major steps in the intellectual development of thinking about self-reproducing and evolving machines:

1. The first step involved the introduction of the view that animals can be understood as machines, due in large part to René Descartes in the 1630s–40s. This step, which we discuss in Chap. 2, introduced—implicitly at first, but later more explicitly—the first glimmerings of the idea of machine self-reproduction. The first direct mentions we find of the idea are in *ab absurdo* arguments against the view of animals as machines, but in the eighteenth and nineteenth centuries we begin to see the idea discussed without necessarily being rejected as obviously absurd. Using the terminology we introduced earlier, the discussion of the subject during this period focused on the possibility of standard self-reproducing machines (*standard-replicators*).

---

[4] Throughout the book we generally use the terms *reproduction* and *replication* synonymously. Within a certain subset of disciplines concerned with self-reproducing machines there is a convention of using the word *replication* in the case where a perfect copy is produced (so there is no evolution) and *reproduction* in the case where mutations and other genetic operators might produce variety in the offspring and thereby allow the possibility of evolution [263]. However, within the broader range of sources that we review here, there are several conflicting definitions of the distinction between these terms (e.g. [74], [128]). Within the context of all of the work we discuss, we believe it is clearer to make the distinction between standard-replicators, evo-replicators and maker-replicators, rather than relying upon the reader to remember technical distinctions between the words reproduction and replication.

2. The second step involved the development of the idea that machines, like animals, might not only be endowed with the capability of *self-reproduction* but also of *evolution* (that is, *evo-replicators*). After the first step had been taken, two further important factors contributed to the realisation of the second step—the climax of the Industrial Revolution in Great Britain in the early nineteenth century, and the publication in 1859 of Darwin's ideas of evolution by natural selection in *On the Origin of Species*. Within a decade of the publication of Darwin's work, we see several authors discuss at length the idea of the evolution of self-reproducing machines. The most significant early work on this topic comes from Samuel Butler, Alfred Marshall and George Eliot, as we discuss in Chap. 3. These works mark the arrival of the concept of evo-replicators in the published literature. Coming as they did at the end of the British Industrial Revolution, these ideas now seemed much less far-fetched, and therefore potentially more frightening. Being a more realistic prospect, this step also led to the development of more thorough discussions of the implications of evo-replicators—in addition to Butler, Marshall and Eliot, John Desmond Bernal in the 1920s provided a deep, scientifically-grounded discussion of how such technology might shape the direction of the far future of humanity. At the same time, evo-replicators became a common theme in early works of science fiction. These early twentieth century developments are covered in Chap. 4.

3. The third step, discussed in Chap. 5, saw the first serious studies of the design and implementation of practical self-reproducing machines. This step comprised two separate strands. One strand was inspired by Alan Turing's work on the concept of *universal computing machines* in the 1930s [293] and involved the development by John von Neumann—starting in the late 1940s—of a theory of *universal constructing machines*. Although von Neumann was very much interested in the potential for self-reproducing machines to evolve, the universal construction aspects of his theory, which concerned general-purpose manufacturing machines, can be seen as the seed of the concept of a *maker-replicator*. During the same period, another strand of work was inspired (in part) by developments in molecular genetics and in understanding the process of DNA self-replication. It involved the study of much simpler artificial self-reproducing systems than those considered in the first strand, and it is exemplified by the first instantiations of artificial self-replicators in software (by Nils Aall Barricelli) and in hardware (by Lionel Penrose). This strand represents the first attempts at implementing *evo-replicators*.

In Chaps. 2–5, we flesh out the detail of these three steps and discuss the people and ideas involved. These chapters take us up to the early 1960s. The period from that point onward is already fairly well documented—in Chap. 6 we provide an overview of this more recent work and give references to existing reviews. We also mention some very recent work that has not been covered elsewhere.

Having traced the development of these ideas, and the thoughts and motivations of those involved, we end by summarising in Chap. 7 what has been achieved, what key issues remain unresolved, the outlook for future work in this area, and the implications for the future of humanity.

But first we need to go back to the beginning. In the next chapter we investigate where, when and why the concept of self-reproducing machines first arose.

Chapter 2
Animals and Machines: Changing Relationships in the 17th & 18th Centuries

We begin our journey by looking at the early intellectual precursors to the idea of self-reproducing machines, dating from the seventeenth and eighteenth centuries. These works form part of a much older exploration of the relationship between living organisms and machines—a tradition with origins in tales of automata from across the ancient world (e.g. [78], [64], [158, pp. 14–22], [202, pp. 31–34]).[1]

2.1 Animals as Machines, Machines as Animals

In the early seventeenth century, René Descartes argued that animals are machines, and that humans alone possess a mind with subjective consciousness:

> ... it seems reasonable since art copies nature, and men can make various automata which move without thought, that nature should produce its own automata much more splendid than the artificial ones. These natural automata are the animals.
>
> René Descartes, *letter to Henry More*, 5 February 1649[2]

Having separated the concepts of life and mind, Descartes' programme for understanding living phenomena envisaged "the reduction of living things to a class of processes which are entirely accessible to general physics" [58, p. 208]. Because he viewed animals as machines, and their reproduction as a purely physical process,[3]

---

[1] More detailed accounts of lifelike automata emerge in medieval Europe and the Middle East, particularly from the thirteenth century onward (e.g. [250], [158]).

[2] Translated text quoted from [69, p. 324].

[3] Descartes' views on reproduction and growth developed over his lifetime [83, pp. 32–33], with the most mature account set out in *De la formation de l'animal* published posthumously alongside *Traité de l'homme* in 1664 [253, p. 4]. He conceived the process as one of regeneration and conservation of form, which followed general laws of nature, progressing from the initial stages of the "extremely volatile and expansive seminal mixture" of the fertilised ovum, with growth constrained and guided by the particular conformation of the ovum's membrane [58, p. 211]. However, his account was ultimately unsatisfying (e.g. [253, pp. 4–5], [83, pp. 33–35], [117, pp. 368–370]). An interesting discussion of Descartes' views in the context of other seventeenth century explana-

© Springer Nature Switzerland AG 2020
T. Taylor, A. Dorin, *Rise of the Self-Replicators*,
https://doi.org/10.1007/978-3-030-48234-3_2

the general idea of self-reproducing machines can be seen as inherent in Descartes' approach. However, his writing about reproductive processes was always with reference to natural, rather than artificial, automata (that is, animals and the human body). Nevertheless, his arguments inevitably led some to pursue the analogy and imagine the idea of self-reproducing artificial automata.

There is an anecdote retold several times in the recent literature about a conversation between Descartes and Queen Christina of Sweden (e.g. [119, 264, 1]).[4] Upon hearing Descartes' views on animals as machines, Christina is said to have responded that "she had never seen her watch give birth to baby watches" [223, p. 19]. We have been unable to find an authoritative original source for this anecdote, and it is probably apocryphal.[5] However, such ideas were certainly in the air by the second half of the seventeenth century; in 1683, for example, the French academic Bernard Le Bovier de Fontenelle wrote:

> Do you say that beasts are machines just as watches are? Put a male dog-machine and a female dog-machine side by side, and eventually a third little machine will be the result, whereas two watches will lie side by side all their lives without ever producing a third watch.
>
>   Bernard Le Bovier de Fontenelle, *Letter XI "to Monsieur C..."*, 1683[6]

While Descartes sought to understand all living phenomena apart from the human mind in terms of material properties, other philosophers of the time adopted a completely materialist perspective, in which *all* phenomena, including the human mind, were to be explained in terms of the properties of matter. One of the most prominent statements of materialism of the period came from the English philosopher Thomas Hobbes; where Descartes had suggested that living beasts are machines, Hobbes in his 1651 book *Leviathan* suggested that mechanical automata possess "an artificial life":

> For seeing life is but a motion of limbs, the beginning whereof is in some principal part within, why may we not say that all automata (engines that move themselves by springs and wheels as doth a watch) have an artificial life? For what is the heart, but a spring; and the nerves, but so many strings; and the joints, but so many wheels, giving motion to the whole body, such as was intended by the artificer?
>
>   Thomas Hobbes, *Leviathan*, 1651 [141, p. 7]

---

tions of biological reproduction can be found in [117], and discussion of further development of thought on these matters in the eighteenth century can be found in [94] and [253].

[4] And [188, p. 16] has "the queen of France" as the protagonist.

[5] In addition to conducting an extensive search of primary and secondary literature, we have also contacted many leading historians of science and philosophy who specialise in Descartes, Queen Christina or early reproductive biology (see acknowledgements at the end of the book), but most of them had not heard of this anecdote, none knew of an original source and most thought it sounded apocryphal. During our search, the earliest mention we found of the anecdote was in a 1924 text in Spanish by the philosopher José Ortega y Gasset [224, p. 610]: "A Descartes, que sostenía la naturaleza mecánica de los cuerpos vivos, ya decía Cristina de Suecia que ¡¡ella no había visto nunca que su reloj diese a luz relojitos¿¿." An English translation of this work appeared in 1941 [223, pp. 18–19]: "Queen Christina of Sweden remarked to Descartes, who upheld the mechanical nature of living beings, that she had never seen her watch give birth to baby watches."

[6] [76, pp. 322–323], translated text quoted from [253, p. 1].

The seventeenth and eighteenth centuries witnessed the creation of progressively more sophisticated automata, accompanied by further materialist comparisons between humans and machines. A chief proponent of these materialist views was Julien Offray de La Mettrie, best known for his work *L'homme machine* (*Man a Machine*) published in 1747 [77]; detailed accounts of his work and of the general intellectual development of such ideas in this period can be found elsewhere (e.g., [250, 158]).[7] By the second half of the eighteenth century, the incredulity shown by de Fontenelle and others in the preceding decades regarding the possibility of machine reproduction had become a little more subdued. In 1769, the French philosopher Denis Diderot pushed the discussion of materialism even further in a series of three dialogues, later published collectively under the title *Le Rêve de d'Alembert* [87]. Comparing the operation of the human mind to the oscillations and resonances of a harpsichord's strings, one of the participants in the first dialogue remarks:

> And so if this sentient and animated harpsichord was now endowed with the faculty of feeding and reproducing itself, it would live and, either on its own or with its female partner, give birth to little harpsichords, living and resonating.
>
> Denis Diderot, *Conversation Between D'Alembert and Diderot*, 1769 [87][8]

Diderot uses this image in the context of proposing a materialist description of the development of a sentient animal from a germ cell; machine reproduction was not the focal point of the passage. Nevertheless, it is perhaps the first published work where the idea of a self-reproducing machine is defended in a dialogue rather than being immediately dismissed as absurd.

2.2 Mechanism and Design

The increasingly common comparison of animals to machines in the seventeenth and eighteenth centuries is also reflected in the growing use of what is now known as the *watchmaker analogy* in arguments for the existence of God.[9] The general form of such arguments is as follows: If you were to study the intricate mechanisms and organisation of a watch and observe how they operate to perform the function of telling the time, you would naturally assume that the watch had been designed by a skilled watchmaker who had the intention of creating an artefact to perform that

---

[7] Descartes' and Le Mettrie's views were by no means universally accepted, however. For detailed coverage of the controversies that continued to rage in early modern philosophy on the relationship between organisms and mechanisms, see, for example, [214, 94, 117, 250].

[8] English translation by Ian Johnston (source: https://web.archive.org/web/20170115114438/http://records.viu.ca/johnstoi/diderot/conversation.htm), quoted with permission. Note that Johnston's translation uses the phrase "little keyboards" at the end of this quotation, which we have replaced with "little harpsichords". The original French version uses the same word "clavecin" (harpsichord) throughout the text. The original French version, together with an account of the history of these dialogues, can be found in [192, pp. 83–98].

[9] But note that the history of such arguments dates back much earlier, at least as far as Cicero in the first century BCE [153].

function; equally, if you consider the immeasurably more complicated mechanisms and organisation of the human body, you should similarly conclude that it has been designed by an ultra-intelligent designer, or God.

In recent times the watchmaker analogy has become most closely associated with William Paley and his 1802 publication *Natural Theology* [228], which we will discuss shortly. Paley is of particular interest in the current context because he raised the idea of a *self-reproducing watch* in his argument. However, we can trace the origins of even this use of the idea of self-reproduction in the watchmaker analogy back to the seventeenth century.[10]

The first printed example of the analogy with an allusion to self-reproduction that we are aware of appears in the English clergyman Thomas Doolittle's 1673 publication *The Young Man's Instructor and the Old Man's Remembrancer* [91].[11] In a chapter of the book entitled "That there is a God", Doolittle sets out the following argument:

> Can any thing that *is not*, work or do any thing? *No.*
>
> Can any thing be before it is? *No.*
>
> But that which doth make any thing is before that which is made? *Yes.*
>
> Then if any creature had made it self, it would have been before it was, and it would have acted when it was nothing: but that is impossible: is it not? *Yes.*
>
> Then every thing that was made, was made by something *that is* and was not made: was it not? *Yes.*
>
> And that *which is*, and was not made, must be God? *Yes.*
>
> When you see a good going Watch, a well tuned musical Instrument, a fair built House, or several Letters in a Printing-house, to be put in due order to make significant Words, you conclude, that there was some Workman that did all these things? *Yes.*
>
> So when you see Sun, Moon, and Stars, the Earth, the Sea, Men, Beasts of the Field, Birds of the Air, being they could not make themselves, you conclude, there must be one that is the first cause of all things that are made, or else they could never have been. ... do you not? *Yes.*
>
> Then there is no more reason you should doubt, whether God is, than whether you are: is there? *No.*
>
> Thomas Doolittle, *The Young Man's Instructor*, 1673 [91, pp. 31–32]

We see here that Doolittle argued that even a self-reproducing organism (a creature that "made itself") must have been originally made by an eternal creator.[12] The

---

[10] A text that bears many similarities to Paley's in the use of the watchmaker analogy was published over eighty years prior to *Natural Theology* by the mathematician Bernard Nieuwentyt in a 1716 publication in Dutch, which was translated into English in 1718 [218]. Some have argued that Paley plagiarised Nieuwentyt's work (see [153, p. 120, pp. 69–71]). However, Nieuwentyt's argument did not include the idea of a self-reproducing watch which was present in Paley's work.

[11] Another early version of the watchmaker argument appears in a lecture preached by John Howe in 1690 [53, p. 1060].

[12] We also see in this quote that Doolittle, with his image of "several Letters in a Printing-house, to be put in due order to make significant Words," is using a version of what is now known as the "tornado in a junkyard" argument, made popular in recent times by Fred Hoyle [148, p. 19], and frequently used by present-day creationists in arguments for the existence of God. John Howe's

argument that self-reproduction does not mitigate the need for an original creator was later made more explicitly by William Paley in his *Natural Theology* [228]. Paley (1743–1805), like Doolittle, was an English clergyman and philosopher, and *Natural Theology* was his final published work, appearing in 1802. In it we also find the first explicit and protracted discussion of the concept of a self-reproducing *artificial* (as opposed to biological) machine.

Paley's discussion of the watchmaker analogy in *Natural Theology* begins in Chapter 1 where he asks us to imagine coming across a stone while walking in the country. If we were asked how it had come to be there, it would be perfectly reasonable, he argued, to assume it had always been there. However, the situation would be different, he suggested, if we came across a watch rather than a stone. By inspecting the watch and discovering it to consist of many intricate parts precisely arranged so as to perform the function of displaying the time, we would hardly conclude that, like the stone, it had lain there forever. We would surely conclude, he argued, that there must have been a skilled craftsman or artificer who "comprehended its construction, and designed its use" [228, p. 8].

In Chapter 2, Paley extends the argument by supposing we discovered that, in addition to telling the time[13]

> ... [the watch] possessed the unexpected property of producing, in the course of its movement, another watch like itself ... That it contained within it a mechanism, a system of parts, a mould for instance, or a complex adjustment of laths, files, and other tools, evidently and separately calculated for this purpose ...
>
> William Paley, *Natural Theology*, 1802 [228, p. 11]

He asked what effect such a discovery would have upon our previous conclusion. His answer was that it would not affect our conclusion, for we would still have to account for the design and construction of the original watch in the series of self-reproducing watches: "There cannot be design without a designer" [228, p. 12].

In his argument, Paley was thinking about watches making *exact* copies of themselves (i.e. self-reproduction without variation, or standard-replicators). He got very close to imagining a progressive series of small changes over a lineage of watches when he asked the further question: in what circumstances *would* the fact that the watch is self-reproducing change our conclusion about its origins? "If the difficulty [of a design requiring a designer] were diminished the further we went back [in the lineage of self-reproducing watches], by going back indefinitely we might exhaust it. And this is the only case to which this sort of reasoning applies" [228, p. 13].

However, Paley did not make the final step of imagining an evolutionary sequence of small changes from simple beginnings. In his watch analogy with self-reproduction but lacking variation "[n]o tendency is perceived, no approach towards a diminution of this necessity [of requiring a designer]" [228, p. 13]. He concluded that, even when considering self-reproduction, "[t]he thing required is the intending

---

publication from around the same time as Doolittle, mentioned in footnote 11 (p. 14), also used a similar image [53, p. 1060]. For a full discussion of the history of such arguments, from antiquity to the present day, see [153].

[13] Note that Paley, like Diderot before him (Sect. 2.1), is using the image of a self-reproducing machine without dismissing it as an obviously absurd or impossible idea.

mind, the adapting hand, the intelligence by which that hand was directed" [228, p. 14]. The step of replacing an intelligently-directed adapting hand with the adapting hand of evolution by natural selection would have to wait another 60 years for the publication of Darwin's *On the Origin of Species* and the imagination of Samuel Butler (see Sect. 3.1).[14]

*   *   *

As we have seen in this chapter, comparisons between animals and machines in the seventeenth and eighteenth centuries led to the first glimmerings of the idea of machine self-reproduction. As time progressed, we see the notion being used in the literature with fewer caveats attached. As outlined in Sect. 1.6, this work represents the first major step in the development of thinking about self-replicators. Specifically, discussions during this period were about the fundamental question of whether a machine could construct a copy of itself—that is, to use our terminology, whether it is possible to design a *standard-replicator*. This step, combined with the introduction and widespread use of increasingly complex machinery during the Industrial Revolution of the late eighteenth and early nineteenth century, and the publication of Darwin's theory of evolution by natural selection, set the stage for the second major step—the idea that machines might not only be able to reproduce but also evolve. The development of this second step—the emergence of the idea of an *evo-replicator*—is the subject of the next two chapters.

---

[14] A tantalising "near miss" in terms of earlier candidates for thinking about the evolution of self-reproducing machines is Erasmus Darwin (1731–1802), a contemporary of William Paley and grandfather of Charles Darwin. He is well known for his early ideas on the theory of biological evolution, set out in *Zoonomia: Or the Laws of Organic Life* [73]. Perhaps less widely known, Erasmus Darwin was also interested in designing machines and mechanisms, examples of which included devices for producing multiple copies of handwritten text and also the design for an artificial mechanical bird, powered by compressed air and featuring a fully specified wing movement cycle [163]. However, despite his combined interests in evolution, mechanical copying machines and artificial organisms, we have found no evidence that he thought about the possibility of self-reproducing machines. Further information and pictures of Darwin's designs are available on the Revolutionary Players website at http://www.revolutionaryplayers.org.uk/the-scope-and-nature-of-darwins-commonplace-book/. A model based upon Darwin's design for an artificial bird was recently commissioned for public display at his former home (now a museum) in Lichfield, England (see http://www.bbc.co.uk/news/uk-england-stoke-staffordshire-21630920).

Chapter 3
Babbage Meets Darwin: Mechanisation and Evolution in the 19th Century

By the climax of the British Industrial Revolution in the 1840s, the idea of machines making other machines was no longer quite such an "unexpected property" as it was when Paley wrote *Natural Theology* just decades earlier. Indeed, around this time we start to see more anxiety about the potential consequences of machine self-reproduction as the idea begins to seem a little less far-fetched. In 1844, the British author and future prime minister Benjamin Disraeli wrote the novel *Coningsby*; in a passage of the book describing the industrial landscape of Manchester, the narrator remarks:

> And why should one say that the machine does not live? It breathes ... It moves ... And has it not a voice? ... And yet the mystery of mysteries is to view machines making machines; a spectacle that fills the mind with curious, and even awful, speculation.
>
> Benjamin Disraeli, *Coningsby*, 1844 [88, p. 154]

By the mid-nineteenth century, the intellectual advances of the preceding two centuries had laid the groundwork for the first extended explorations of the possible repercussions of self-reproducing machines, with particular concern about their potential to *evolve*. The nexus of this development was London; in the nineteenth century, the intellectual elite of England were a richly connected web of thinkers, among whom ideas of science, philosophy, technology, literature and the arts freely flowed.

For example, shortly after returning from his voyage on the *Beagle*, Charles Darwin attended one of Charles Babbage's regular London soirées and witnessed a demonstration of his work on a mechanical calculating machine, the *Difference Engine*.[1] Babbage worked with Augusta Ada King, Countess of Lovelace (Ada Lovelace), who was the daughter of poet George (Lord) Byron, whose friend Percy

---

[1] Babbage used the Difference Engine to demonstrate how discontinuities could arise in a system without external intervention, and thereby to argue that discontinuities in Nature, such as the appearance of new species, could likewise be explained by natural laws without requiring constant divine guidance [267]. Snyder argues that this demonstration likely emboldened Darwin's ideas of nature being governed by natural laws [267, p. 195]. Babbage's fascination with complex machines had been stimulated as a young child when he was taken to exhibitions of human and animal mechanical automata [250, p. 126–127], [202, p. 134–135].

© Springer Nature Switzerland AG 2020
T. Taylor, A. Dorin, *Rise of the Self-Replicators*,
https://doi.org/10.1007/978-3-030-48234-3_3

Bysshe Shelley was married to novelist Mary Shelley, whose work *Frankenstein* [261] remains the seminal science fiction account of the creation of an artificial being,[2] and so on ...

It was into this intellectual powder keg that Darwin was about to drop a lit match with the publication of *On the Origin of Species* in 1859 [72]. Descartes and La Mettrie had claimed that organisms were machines, and Darwin now argued that complex organisms had evolved from simple beginnings. We do not have to wait long thereafter to find thinkers who combined these lines of thought to conceive and explore the idea of self-reproducing, evolving machines—that is, *evo-replicators*.

Within a year of Darwin's publication, his American friend and colleague Asa Gray published an extended review and examination of Darwin's theory in three successive issues of *The Atlantic* magazine.[3] In the first of these, published in July 1860, Gray compared Darwin's picture of biological evolution to the development of human technology and artefacts:

> To compare small things with great in a homely illustration: man alters from time to time his instruments or machines, as new circumstances or conditions may require and his wit suggest. Minor alterations and improvements he adds to the machine he possesses; he adapts a new rig or a new rudder to an old boat: this answers to *variation*. If boats could engender, the variations would doubtless be propagated, like those of domestic cattle. In course of time the old ones would be worn out or wrecked; the best sorts would be chosen for each particular use, and further improved upon; and so the primordial boat be developed into the scow, the skiff, the sloop, and other species of water-craft,—the very diversification, as well as the successive improvements, entailing the disappearance of many intermediate forms, less adapted to any one particular purpose; wherefore these go slowly out of use, and become extinct species: this is *natural selection*.
>
> Asa Gray, 1860 [126, p. 122] (original emphasis)

This passage is perhaps the first published example of an analogy being drawn between the process of biological evolution and the evolution of human technology.[4]

---

[2] For a detailed discussion of the scientific context (especially the biological and evolutionary context) within which Shelley wrote Frankenstein, see [227]. Shelley's novel itself includes an allusion to self-reproducing artificial beings, if not automata as such. In a section where Victor Frankenstein has commenced making a female companion for his creature, he decides to halt the endeavour after imagining what might come of it: "Even if they were to leave Europe, and inhabit the deserts of the new world, yet one of the first results of those sympathies for which the daemon thirsted would be children, and a race of devils would be propagated upon the earth, who might make the very existence of the species of man a condition precarious and full of terror" [261, ch. 20] (see also [31, p. 195]).

[3] These articles were later collated as a single volume and published in London the following year [127].

[4] George Basalla's classic book *The Evolution of Technology* does not mention Gray's work, but instead focuses on organic-mechanical analogies in the works of Samuel Butler (whom we describe in Sect. 3.1) and Augustus Pitt-Rivers (born Augustus Lane-Fox) [21, pp. 14–21]. It is true that Butler and Pitt-Rivers explored these ideas far more than did Gray. Basalla identifies a variety of other writers who further developed the idea of the evolution of technology in the late nineteenth and early twentieth centuries, but to elaborate on those here would take us beyond our focus on *self-reproducing* machines. Here we mention just one further work of interest, which is little cited elsewhere: in 1910 the German chemist and Nobel laureate Wilhelm Ostwald published a short article in the supplement of *Scientific American* entitled "*Machines and Living Creatures*"

In addition, Gray also conjures the image of a boat giving rise to other boats like itself. While Gray was not necessarily thinking of *self*-reproducing technology,[5] over the following 20 year period we find at least three authors writing explicitly and extensively about the idea of evo-replicators—and about the potential consequences of their emergence for the future of humankind.

3.1 Samuel Butler: *Darwin Among the Machines* (1863)

A little biographical background is required to explain how, in the 1860s, a sheep farmer on a remote ranch in New Zealand came to be writing about the conquest of humanity by evolving, self-reproducing machines.

Born in England, Samuel Butler (1835–1902) set sail aboard the *Roman Emperor*, bound for New Zealand, in October 1859 [44].[6] A recent graduate of Cambridge University, and son of the Reverend Thomas Butler,[7] Samuel had decided not to follow his father into the clergy, but instead intended to establish himself among the early British settlers of the Canterbury settlement in New Zealand's South Island.

Butler's plan was to increase his wealth by sheep farming, and by late 1860 he had established a sheep run named Mesopotamia Station. Although the run was situated some 90 miles south-west of the regional capital Christchurch, Butler spent much of his time in the capital and became well connected. Over the four years he spent in New Zealand, he wrote a number of contributions for *The Press*, a Christchurch newspaper.

Shortly after his arrival in New Zealand, Butler obtained a copy of Darwin's recently-published *The Origin of Species* [72]. In late 1862 he anonymously published a dialogue in *The Press* entitled *Darwin on the Origin of Species* [42] in which one character argues in support of Darwin's theory and the other against it. This wrestling with Darwin's theory and its implications was to form a common

---

[225]. Ostwald argued that machines and biological organisms can both be considered as energy transformers, and both evolve over time to achieve "a higher ratio of efficiency between energy consumed and energy produced." Both the removal of dispensable parts, and the development of individual organs to perform specific functions, can be explained by this principle in organisms and in machines, said Ostwald. The only subsequent reference we have found to Ostwald's article appears in an editorial of an American Theosophy newsletter, *Century Path*, later in the same year [287]—this criticised the application of analogies of biological evolution to human technology on the grounds that the latter requires conscious human selection for its operation.

[5] The word "engender" in his phrase "If boats could engender ..." is ambiguous in this regard. Interestingly, the 1861 republication of the work includes an expanded version of this sentence, which reads "'Like begets like,' being the great rule in nature, if boats could engender, the variations would doubtless be propagated, like those of domestic cattle." [127, p. 6].

[6] Sources of biographical details in this section include [146], [252] and [48], and further information is also available in [97].

[7] Samuel Butler was also the grandson of Dr Samuel Butler, headmaster of Shrewsbury School during the years when the young Charles Darwin attended the school [114]. Furthermore, he was distantly related, by marriage, to William Paley [124, p. 12].

thread in many of Butler's later works. The following year, on 13 June 1863, he published another letter in *The Press*—this time under the pseudonym *Cellarius*—entitled *Darwin Among the Machines* [43] (see Fig. 3.1).

Butler began the letter by noting the rapid pace of development of machinery from the earliest mechanisms to the most sophisticated examples of the day. He commented that this had far outstripped the pace of development in the animal and vegetable kingdoms, and asked what might be the ultimate outcome of this trend. Observing the increasingly sophisticated "self-regulating, self-acting power" with which machines were being conferred, Butler suggested that humans "are ourselves creating our own successors" [43, p. 1]. He further speculated that, freed from the constraints of feelings and emotion, machines will ultimately become "the acme of all that the best and wisest man can ever dare to aim at," at which point "man will have become to the machine what the horse and the dog are to man" [43, p. 1].[8]

At that stage, Butler reasoned, the machines would still be reliant upon humans for feeding them, repairing them and producing their offspring, and hence they would likely treat us kindly. "[Man] will continue to exist, nay even to improve, and will be probably better off in his state of domestication under the beneficent rule of the machines than he is in his present wild state" [43, p. 1]. However, he then introduced the possibility of a time when "the reproductive organs of the machines have been developed in a manner which we are hardly yet able to conceive," noting that "it is true that machinery is even at this present time employed in begetting machinery, in becoming the parent of machines often after its own kind" [43, p. 2].

Given the scenario he had presented, Butler ended the letter by suggesting that the best course of action for the human race was to embark upon the destruction of all machines and to return to a simpler way of life. If such a course of action seemed impossible given the degree to which human civilisation already relied upon technology, Butler warned that "the mischief is already done, that our servitude has commenced in good earnest, that we have raised a race of beings whom it is beyond our power to destroy, and that we are not only enslaved but are absolutely acquiescent in our bondage" [43, p. 2].

Among subsequent commentators on Butler's work, there are varying opinions on whether his intention was to support or critique Darwin's theory (see [146, pp.

---

[8] We can see glimmerings of the modern idea of the *technological singularity* [209, 175] (sometimes simply referred to as "the singularity") in Butler's writing. Different authors adopt different definitions of this concept, but they all essentially involve a profound change in human civilisation brought about by the emergence of machines with greater-than-human intelligence [99, pp. 1–12]. The differences lie in the rate at which these changes might happen. Nick Bostrom has recently introduced the term *superintelligence* as a more precise concept that does not commit to the ongoing exponential growth rates envisaged by some proponents of the singularity [29]. As we see here, the origin of these ideas can be traced back at least as far as Butler's writing in the 1860s. The birth of the digital computer age prompted increasing interest and speculation along these lines; the idea was discussed in the 1950s by both Alan Turing [292, p. 666] (who explicitly referred to Butler's work) and John von Neumann [296, p. 5] (who first introduced the term *singularity* in reference to the accelerating pace of technology, and whose work on machine self-reproduction we discuss in Sect. 5.1.1). For a good discussion of the history of these ideas and an analysis of the themes entailed, see [99], and for a recent general overview see [259].

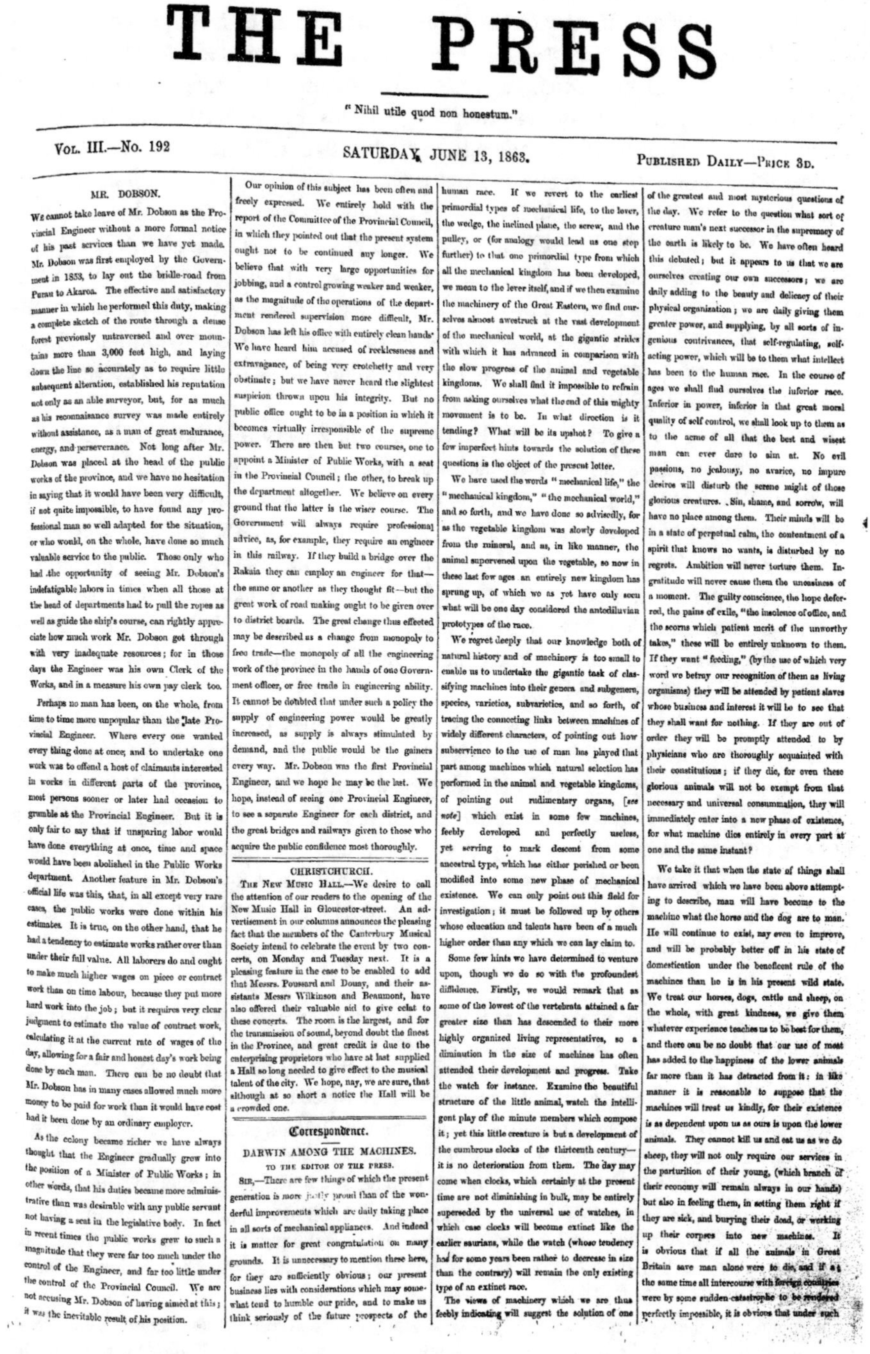

THE PRESS

"Nihil utile quod non honestum."

Vol. III.—No. 192 SATURDAY, JUNE 13, 1863. Published Daily—Price 3d.

MR. DOBSON.

We cannot take leave of Mr. Dobson as the Provincial Engineer without a more formal notice of his past services than we have yet made. Mr. Dobson was first employed by the Government in 1853, to lay out the bridle-road from Purau to Akaroa. The effective and satisfactory manner in which he performed this duty, making a complete sketch of the route through a dense forest previously untraversed and over mountains more than 3,000 feet high, and laying down the line so accurately as to require little subsequent alteration, established his reputation not only as an able surveyor, but, for as much as his reconnaisance survey was made entirely without assistance, as a man of great endurance, energy, and perseverance. Not long after Mr. Dobson was placed at the head of the public works of the province, and we have no hesitation in saying that it would have been very difficult, if not quite impossible, to have found any professional man so well adapted for the situation, or who would, on the whole, have done so much valuable service to the public. Those only who had the opportunity of seeing Mr. Dobson's indefatigable labors in times when all those at the head of departments had to pull the ropes as well as guide the ship's course, can rightly appreciate how much work Mr. Dobson got through with very inadequate resources; for in those days the Engineer was his own Clerk of the Works, and in a measure his own pay clerk too.

Perhaps no man has been, on the whole, from time to time more unpopular than the late Provincial Engineer. Where every one wanted every thing done at once, and to undertake one work was to offend a host of claimants interested in works in different parts of the province, most persons sooner or later had occasion to grumble at the Provincial Engineer. But it is only fair to say that if unsparing labor would have done everything at once, time and space would have been abolished in the Public Works department. Another feature in Mr. Dobson's official life was this, that, in all except very rare cases, the public works were done within his estimates. It is true, on the other hand, that he had a tendency to estimate works rather over than under their full value. All laborers do and ought to make much higher wages on piece or contract work than on time labour, because they put more hard work into the job; but it requires very clear judgment to estimate the value of contract work, calculating it at the current rate of wages of the day, allowing for a fair and honest day's work being done by each man. There can be no doubt that Mr. Dobson has in many cases allowed much more money to be paid for work than it would have cost had it been done by an ordinary employer.

As the colony became richer we have always thought that the Engineer gradually grew into the position of a Minister of Public Works; in other words, that his duties became more administrative than was desirable with any public servant not having a seat in the legislative body. In fact in recent times the public works grew to such a magnitude that they were far too much under the control of the Engineer, and far too little under the control of the Provincial Council. We are not accusing Mr. Dobson of having aimed at this; it was the inevitable result of his position.

Our opinion of this subject has been often and freely expressed. We entirely hold with the report of the Committee of the Provincial Council, in which they pointed out that the present system ought not to be continued any longer. We believe that with very large opportunities for jobbing, and a control growing weaker and weaker, as the magnitude of the operations of the department rendered supervision more difficult, Mr. Dobson has left his office with entirely clean hands. We have heard him accused of recklessness and extravagance, of being very crotchetty and very obstinate; but we have never heard the slightest suspicion thrown upon his integrity. But no public office ought to be in a position in which it becomes virtually irresponsible of the supreme power. There are then but two courses, one to appoint a Minister of Public Works, with a seat in the Provincial Council; the other, to break up the department altogether. We believe on every ground that the latter is the wiser course. The Government will always require professional advice, as, for example, they require an engineer in this railway. If they build a bridge over the Rakaia they can employ an engineer for that—the same or another as they thought fit—but the great work of road making ought to be given over to district boards. The great change thus effected may be described as a change from monopoly to free trade—the monopoly of all the engineering work of the province in the hands of one Government officer, or free trade in engineering ability. It cannot be doubted that under such a policy the supply of engineering power would be greatly increased, as supply is always stimulated by demand, and the public would be the gainers every way. Mr. Dobson was the first Provincial Engineer, and we hope he may be the last. We hope, instead of seeing one Provincial Engineer, to see a separate Engineer for each district, and the great bridges and railways given to those who acquire the public confidence most thoroughly.

CHRISTCHURCH.

The New Music Hall.—We desire to call the attention of our readers to the opening of the New Music Hall in Gloucester-street. An advertisement in our columns announces the pleasing fact that the members of the Canterbury Musical Society intend to celebrate the event by two concerts, on Monday and Tuesday next. It is a pleasing feature in the case to be enabled to add that Messrs. Poussard and Douay, and their assistants Messrs Wilkinson and Beaumont, have also offered their valuable aid to give eclat to these concerts. The room is the largest, and for the transmission of sound, beyond doubt the finest in the Province, and great credit is due to the enterprising proprietors who have at last supplied a Hall so long needed to give effect to the musical talent of the city. We hope, nay, we are sure, that although at so short a notice the Hall will be a crowded one.

Correspondence.

DARWIN AMONG THE MACHINES.

TO THE EDITOR OF THE PRESS.

Sir,—There are few things of which the present generation is more justly proud than of the wonderful improvements which are daily taking place in all sorts of mechanical appliances. And indeed it is matter for great congratulation on many grounds. It is unnecessary to mention these here, for they are sufficiently obvious; our present business lies with considerations which may somewhat tend to humble our pride, and to make us think seriously of the future prospects of the human race. If we revert to the earliest primordial types of mechanical life, to the lever, the wedge, the inclined plane, the screw, and the pulley, or (for analogy would lead us one step further) to that one primordial type from which all the mechanical kingdom has been developed, we mean to the lever itself, and if we then examine the machinery of the Great Eastern, we find ourselves almost awestruck at the vast development of the mechanical world, at the gigantic strides with which it has advanced in comparison with the slow progress of the animal and vegetable kingdoms. We shall find it impossible to refrain from asking ourselves what the end of this mighty movement is to be. In what direction is it tending? What will be its upshot? To give a few imperfect hints towards the solution of these questions is the object of the present letter.

We have used the words "mechanical life," the "mechanical kingdom," "the mechanical world," and so forth, and we have done so advisedly, for as the vegetable kingdom was slowly developed from the mineral, and as, in like manner, the animal supervened upon the vegetable, so now in these last few ages an entirely new kingdom has sprung up, of which we as yet have only seen what will be one day considered the antediluvian prototypes of the race.

We regret deeply that our knowledge both of natural history and of machinery is too small to enable us to undertake the gigantic task of classifying machines into their genera and subgenera, species, varieties, subvarieties, and so forth, of tracing the connecting links between machines of widely different characters, of pointing out how subservience to the use of man has played that part among machines which natural selection has performed in the animal and vegetable kingdoms, of pointing out rudimentary organs, [*see note*] which exist in some few machines, feebly developed and perfectly useless, yet serving to mark descent from some ancestral type, which has either perished or been modified into some new phase of mechanical existence. We can only point out this field for investigation; it must be followed up by others whose education and talents have been of a much higher order than any which we can lay claim to.

Some few hints we have determined to venture upon, though we do so with the profoundest diffidence. Firstly, we would remark that as some of the lowest of the vertebrata attained a far greater size than has descended to their more highly organized living representatives, so a diminution in the size of machines has often attended their development and progress. Take the watch for instance. Examine the beautiful structure of the little animal, watch the intelligent play of the minute members which compose it; yet this little creature is but a development of the cumbrous clocks of the thirteenth century—it is no deterioration from them. The day may come when clocks, which certainly at the present time are not diminishing in bulk, may be entirely superseded by the universal use of watches, in which case clocks will become extinct like the earlier saurians, while the watch (whose tendency has for some years been rather to decrease in size than the contrary) will remain the only existing type of an extinct race.

The views of machinery which we are thus feebly indicating will suggest the solution of one of the greatest and most mysterious questions of the day. We refer to the question what sort of creature man's next successor in the supremacy of the earth is likely to be. We have often heard this debated; but it appears to us that we are ourselves creating our own successors; we are daily adding to the beauty and delicacy of their physical organization; we are daily giving them greater power, and supplying, by all sorts of ingenious contrivances, that self-regulating, self-acting power, which will be to them what intellect has been to the human race. In the course of ages we shall find ourselves the inferior race. Inferior in power, inferior in that great moral quality of self control, we shall look up to them as to the acme of all that the best and wisest man can ever dare to aim at. No evil passions, no jealousy, no avarice, no impure desires will disturb the serene might of those glorious creatures. Sin, shame, and sorrow, will have no place among them. Their minds will be in a state of perpetual calm, the contentment of a spirit that knows no wants, is disturbed by no regrets. Ambition will never torture them. Ingratitude will never cause them the uneasiness of a moment. The guilty conscience, the hope deferred, the pains of exile, "the insolence of office, and the scorns which patient merit of the unworthy takes," these will be entirely unknown to them. If they want "feeding," (by the use of which very word we betray our recognition of them as living organisms) they will be attended by patient slaves whose business and interest it will be to see that they shall want for nothing. If they are out of order they will be promptly attended to by physicians who are thoroughly acquainted with their constitutions; if they die, for even these glorious animals will not be exempt from that necessary and universal consummation, they will immediately enter into a new phase of existence, for what machine dies entirely in every part at one and the same instant?

We take it that when the state of things shall have arrived which we have been above attempting to describe, man will have become to the machine what the horse and the dog are to man. He will continue to exist, nay even to improve, and will be probably better off in his state of domestication under the beneficent rule of the machines than he is in his present wild state. We treat our horses, dogs, cattle and sheep, on the whole, with great kindness, we give them whatever experience teaches us to be best for them, and there can be no doubt that our use of meat has added to the happiness of the lower animals far more than it has detracted from it: in like manner it is reasonable to suppose that the machines will treat us kindly, for their existence is as dependent upon us as ours is upon the lower animals. They cannot kill us and eat us as we do sheep, they will not only require our services in the parturition of their young, (which branch of their economy will remain always in our hands) but also in feeding them, in setting them right if they are sick, and burying their dead, or working up their corpses into new machines. It is obvious that if all the animals in Great Britain save man alone were to die, and if at the same time all intercourse with foreign countries were by some sudden catastrophe to be rendered perfectly impossible, it is obvious that under such

Fig. 3.1 The front page of *The Press* newspaper from 13 June 1863. Butler's letter *Darwin Among The Machines* starts at the bottom of the second column. The letter continued to the end of the first column of the second page (not shown).

26–28]).[9] Butler was certainly challenged by the implications of the theory, and his published work reflects his own conflicting views and their development over his lifetime. For example, in a subsequent letter to *The Press* entitled *Lucubratio Ebria* [45], published on 29 July 1865, he presented a vision of machines as an extension of the human body, rather than as a competing species. From this perspective, Butler emphasised the capacity of machines to exert positive influences on the evolution of humankind, not only by increasing our physical and mental capabilities but also by changing the environment in which we develop as individuals and evolve as a species.[10] He wrote: "We are children of the plough, the spade, and the ship; we are children of the extended liberty and knowledge which the printing press has diffused. Our ancestors added these things to their previously existing members; the new limbs were preserved by natural selection and incorporated into human society; they descended with modifications, and hence proceeds the difference between our ancestors and ourselves." [45].

Butler's struggle in deciding how best to reconcile Darwin's theory of biological evolution with its implications for machine evolution is well expressed in *Lucubratio Ebria*: "We know that what we see is but a sort of intellectual Siamese twins, of which the one is substance and the other shadow, but we cannot set either free without killing both." Upon his return to England in 1864, he continued to explore these ideas, first in an expanded essay entitled *The Mechanical Creation*, published in the London journal *The Reasoner* in 1865 [46],[11] and finally in their most developed form as *The Book of the Machines*, which constituted chapters 23–25 of his novel *Erewhon*, published in 1872 [47].

In *Erewhon*, Butler explored the idea of the collective reproduction of heterogeneous groups of machines, as an alternative to a single machine individually producing a complete copy of itself. He likened a complicated machine to "a city or society" [47, p. 212], and asked "how few of the machines are there which have not been produced systematically by other machines?" [47, p. 210].[12] He invoked a number of biological analogies, such as bee pollination and specialisation of reproductive function in ant colonies, to argue that machines are no less lifelike even if not fully *self*-reproducing individually.

---

[9] The ambiguity is partially attributable to Butler's literary style. E. M. Forster, whose short story *The Machine Stops* we discuss in Sect. 4.1.1, was influenced both by Butler's ideas and his technique [116]. He praised Butler as "a master of the oblique" whose technique involved "muddling up the actual and the impossible until the reader isn't sure which is which."

[10] To apply modern terminology from theoretical biology to Butler's ideas, we could say that he is arguing that machines can be viewed as part of the *extended phenotype* of humans [75], and that human evolution is affected by *niche construction* through our machine-building activities [219].

[11] Butler's *Lucubratio Ebria* (LE) was also written after his return to England, and he sent it from London to New Zealand to be published in *The Press*. Its eventual publication date of 29 July 1865 was a few weeks after that of *The Mechanical Creation* (MC) on 1 July 1865, although Butler wrote LE before MC [202, p. 150].

[12] These and all other quotes from *Erewhon* in this section are voiced by characters in the novel. However, as this part of the novel was a development of Butler's earlier thoughts set out in *The Mechanical Creation* and *Darwin Among the Machines* (as explained in the preface of *Erewhon* [47, p. 33]), we can assume that these quotes are representative of Butler's own thinking.

He further explored the idea, first addressed in *Lucubratio Ebria*, that humans and machines are *co*-evolving, in a process driven by market economics. However, in contrast to his earlier writing, he now feared that this might be detrimental to humankind, with machines evolving to act parasitically upon their designers: "[the machines] have preyed upon man's grovelling preference for his material over his spiritual interests" [47, p. 207]. Humans, he argued, are economically invested in producing machines with ever more "intelligibly organised" mechanical reproductive systems [47, p. 212]:

> For man at present believes that his interest lies in that direction; he spends an incalculable amount of labour and time and thought in making machines breed better and better ... and there seem no limits to the results of accumulated improvements if they are allowed to descend with modification from generation to generation.
>
> Samuel Butler, *Erewhon*, 1872 [47, p. 212]

As machines evolved to become ever more complex, Butler cautioned that they might "so equalise men's powers" that evolutionary selection pressure on human physical capabilities would be reduced to a level that precipitated "a degeneracy of the human race, and indeed that the whole body might become purely rudimentary" [47, p. 224]. This concern about the consequences for the human race of entering a long-term co-evolutionary relationship with machines is taken up by a number of later authors, most notably J. D. Bernal, whose 1929 work *The World, The Flesh and the Devil* we discuss in Sect. 4.2.1.

3.2 Alfred Marshall: *Ye Machine* (c. 1867)

In 1865, the year after Butler's return to London, and a little over 50 miles away, Alfred Marshall (1842–1924)—a recently appointed fellow at St. John's College, Cambridge—was introduced to the university's *Discussion Society* [241]. The society was a forum for intellectual debate that later became known as *The Grote Club* after its founder, the Reverend John Grote.

A graduate in mathematics, Marshall was later to become one of the founding fathers of neoclassical economics. In his influential book *The Principles of Economics*, first published in 1890, Marshall drew analogies between economics and biology. He noted that both dealt with systems "of which the inner nature and constitution, as well as the outer form, are constantly changing" [198, p. 772], and, further, that the development of both biological and industrial organisations "involves an increasing subdivision of functions between its separate parts on the one hand, and on the other a more intimate connection between them" [198, p. 241]. He argued that "[t]he Mecca of the economist lies in economic biology" [198, p. xiv].[13]

---

[13] Several authors have commented, however, that Marshall's plans to develop a more biological economics remained largely aspirational, and a planned sequel to *The Principles of Economics* dealing with economic dynamics was never completed [143, 286].

During his early career in the 1860s, however, Marshall was more engaged in questions of philosophy—in particular, the extent to which the activities of the human mind could be understood in purely physical terms [241, 66].[14] He wrote a series of four papers that formed the basis of talks presented at The Grote Club: "*The Law of Parcimony*", "*Ferrier's Proposition One*",[15] "*Ye Machine*", and "*The Duty of the Logician or the System-maker to the Metaphysician and to the Practical Man of Science*" [241, 242].[16] The first three of these are a sequential discourse as Marshall feels his way "towards a general theory of psychology ... as a doctrine that the action of Brutes are accountable for by mechanical agencies only ... and that the phenomena of the human mind are accountable for by mechanical agencies and self-consciousness" [242, p. 111].

In the first two papers, Marshall defended the distinction between the subjective and objective aspects of the mind. Although supportive of the philosophical work of Ferrier and his followers, he complained that they made no attempt to engage with contemporary scientific approaches to mental phenomena [242, pp. 110–111]. In the third paper, "*Ye Machine*", Marshall addressed this failure by proposing a model for the objective study of mechanisms capable of learning and intelligent action. Inspired by recent scientific work in psychology,[17] evolution[18] and calculating machines,[19] he described the design of a mechanical device (a robot in today's terms) equipped with sensors, effectors and inner circuitry that would allow it to develop progressively more sophisticated ideas and reasoning about its interactions with the world.

Marshall's proposal for the device's "brain" was as follows:

> We may suppose the Machine to contain an indefinite number of wheels of various sizes, and in various positions ... Now suppose that when any two wheels were together performing two partial revolutions, the Machine itself connects them by a light band, slightly fitting. Then, when one of them again revolved, the other would also revolve, unless there were a resisting or opposing force, in which case the band would slip. But every time the same double motion was repeated the band would be tightened.
>
> Alfred Marshall, *Ye Machine*, c. 1867 [242, p. 116]

The proposal was clearly intended as a thought experiment rather than a practical design. But Marshall's essential point was that it is possible to conceive of a machine

---

[14] Indeed, Marshall apparently always regarded himself as a philosopher, writing towards the end of his life that "I taught economics ... but repelled with indignation the suggestion that I was an economist: 'I am a philosopher, straying into a foreign land. I will go home soon.' " [241, p. 53].

[15] James Frederick Ferrier was a nineteenth century Scottish moral philosopher whose Idealist philosophy saw self-consciousness as the defining feature of human experience [241, p. 64]. Marshall describes Ferrier's Proposition One as "along with whatever any intelligence knows, it must, as the ground or condition of its knowledge, have some cognizance of itself" [242, p. 105].

[16] Marshall presented the first of these papers to The Grote Club on 27 March 1867 [241, p. 62]. The precise dates of the subsequent presentations are unknown, although it appears that the second and third papers were presented in two consecutive weeks [242, pp. 111–113].

[17] Most notably Alexander Bain's work on associationism.

[18] Primarily the work of Herbert Spencer rather than Charles Darwin (see [143] for further discussion).

[19] The work of Charles Babbage (see [66] for further discussion).

with a mechanism that strengthens the linkage between internal elements that tend to be active concurrently.[20] The design implements what would now be classified as a kind of *associative learning*. Marshall goes on to describe how a machine like this might also learn through receiving positive or negative feedback about its actions, and how it might develop instincts to maintain desired states. Although such instincts could arise from the associative learning mechanisms already mentioned, Marshall speculated:

> Nay, further, the Machine, like Paley's watch, might make others like itself. We thus get hereditary and accumulated instinct. For these descendants, as they may be called, may vary slightly, owing to accidental circumstances, from the parent. Those which were most suited to the environment would supply themselves most easily with fuel, etc. and have the greatest chance of prolonged activity. The principle of natural selection, which indeed involves only purely mechanical agencies, would thus be in full operation.
>
> Alfred Marshall, *Ye Machine*, c. 1867 [242, p. 119]

He went on to discuss how the Machine's design might be augmented with a second level of inner circuitry, which he called its *Cerebrum* in contrast to its existing *Cerebellum*. Marshall discussed how the Cerebrum could give the Machine the power to reason about sequences of future events by internal meditation. He then addressed its ability to learn concepts of language, numbers and arithmetic, geometry, mechanics and the natural sciences. In his discussion, Marshall invoked the idea of natural selection on a couple of further occasions. He suggested it might assist Machines to evolve complex capacities for cooperation and for the communication of ideas [242, p. 124], along with strong powers of sympathy and moral character [242, p. 130].[21]

"*Ye Machine*" and the other papers presented by Marshall at The Grote Club in the late 1860s had a very limited—albeit distinguished[22]—audience at the time, and they were not published in the scientific literature until 1994 (courtesy of the efforts of the late Tiziano Raffaelli). However, the ideas that Marshall developed in these papers are clear antecedents of themes in his influential work in economics later in his career (discussed earlier in this section).[23]

On a final biographical note, Alfred Marshall's marriage in 1877 provides another example of the interconnectedness of the key figures in our story; his bride was Mary Paley [68], great-granddaughter of none other than William Paley (Sect. 2.2).

---

[20] Marshall also suggested that the Machine could be designed with electromagnetic components [242, pp. 116–117].

[21] Marshall also foresaw—like Butler before him (footnote 8, p. 20)—that the process could lead to machines with superintelligence beyond the level of humans: "a Machine of very great power—by means of the enormous number of associations which it would have ever present with it—might … discover laws that we have not yet attained to, and might set to work to dig for its own coal in places where coal was never heard of" [242, p. 129].

[22] Other active members of The Grote Club at the time included the economist and philosopher Henry Sidgwick, the logician John Venn and the theologian Frederick Denison Maurice [242, p. 103].

[23] See also [143] for further discussion of the role of biological analogies in Marshall's work in economics.

3.3 George Eliot: *Impressions of Theophrastus Such* (1879)

In the following decade, the British author Mary Ann Evans (known by her pen name George Eliot) published her final work *Impressions of Theophrastus Such* [102]. The work is written as a series of short essays by an imaginary scholar named Theophrastus, who attempts to study the human species by focusing on how certain individuals behave in particular interactions—much as an ethologist might study animal behaviour. As the literary scholar S. Pearl Brilmyer puts it, by "[a]massing descriptions of various unperceptive and unsympathetic human beings ... Theophrastus tries to illuminate that which escapes his embodied awareness: the form of the species of which he is but an instance" [34, p. 37].

The penultimate chapter, entitled *Shadows of The Coming Race*, covers in a few short pages many themes concerning the long-term future of the human species. The chapter is written as a dialogue between a character named Trost, who has an optimistic view of the future of humanity, and the narrator, who is more pessimistic. The discussion focuses upon our relationship with machines, and the question of whether consciousness is an advantage to our species or rather an "idle parasite" which we would do better without.

Echoing views expressed in Butler's works, the chapter's narrator foresees a time when machines develop "conditions of self-supply, self-repair, and reproduction." These developments, she fears, will have detrimental effects upon society, leading to mass unemployment. This would first affect the ranks of the lower-skilled, but eventually even the most highly-skilled and intellectual of our species will become redundant and "subside like the flame of a candle in the sunlight," superseded by the machines that are "free from the fussy accompaniment of ... consciousness." Like Butler, Eliot imagined that along the path to our eventual extinction, our increasing reliance upon machines would lead to a degeneration of the human body, leaving us "pale, pulpy, and cretinous."

Many of the ideas expressed in Eliot's chapter can also be found in Butler's earlier works. Indeed, Butler thought that Eliot had "cribbed" *Erewhon* in her work.[24] However, it is perfectly possible that both authors seized upon similar ideas independently, given the intellectual atmosphere of the time as described at the start of this chapter. Eliot was well informed of contemporary scientific developments, and had read *On the Origin of Species* days after its publication [164, pp. 28–30]. Indeed, from the early 1850s she had a close friendship with the philosopher Herbert Spencer, who was an early advocate of the theory of evolution [131, p. 112], and in the late 1860s she became friends with Charles Darwin and his family too.[25] Furthermore, her partner George Henry Lewes had started working on his major work

---

[24] See entry entitled "George Eliot" in Butler's notebooks [49, p. 90] and a letter from Butler dated 10 June 1880 [50, pp. 85–86]. However, the idea might have been originally planted in Butler's mind by his confidant Eliza Savage (see letter from Savage to Butler dated 24 September 1879, [51, pp. 208–210]), and her intentions in doing so might have been somewhat convoluted [164, pp. 224–226].

[25] As documented in various letters between Darwin, Eliot and Eliot's partner George Henry Lewes (source: https://www.darwinproject.ac.uk/george-eliot-mary-ann-evans).

Problems of Life and Mind in 1867, just two years before Eliot wrote *Impressions* [65, p. 463]. Lewes' work included a section entitled *Animal Automatism*, which explored the questions of "whether animals are machines" and "in what sense can we correctly speak of Feeling as an agent in organic processes?" [189, p. 362].[26] There is therefore no doubt that Eliot was immersed in just the scientific and philosophical ideas that would have allowed her to conceive her chapter *Shadows of the Coming Race* completely independently of Butler's work.[27]

There is a fear expressed in Eliot's chapter that scientists are proceeding with a blinkered view of what might be the long-term consequences of their creations. Towards the end of the chapter the narrator summarises her vision of the ultimate outcome of the process of technological development—when humans have been driven to extinction by intelligent but unconscious machines—in the following passage:

> Thus this planet may be filled with beings who will be blind and deaf as the inmost rock, yet will execute changes as delicate and complicated as those of human language and all the intricate web of what we call its effects, without sensitive impression, without sensitive impulse: there may be, let us say, mute orations, mute rhapsodies, mute discussions, and no consciousness there even to enjoy the silence.
>
> George Eliot, *Impressions of Theophrastus Such*, 1879 [102, ch. 17]

Asked where these ideas had come from, the narrator explains that "[t]hey seem to be flying around in the air with other germs." By the late 1800s these topics were indeed very much in the air.

---

[26] These are similar topics to those addressed by Alfred Marshall in his Grote Club lectures, described in Sect. 3.2.

[27] On the other hand, it is also possible, and perhaps likely, that Eliot was making an implicit reference to *Erewhon* in the chapter—this would tie in with her general style that integrates "literary, scientific, or historical allusion in the structure of *Impressions*" [103, p. ix]. Indeed, the chapter title *Shadows of the Coming Race* is a reference to Eliot's friend Edward Bulwer-Lytton's 1871 sci-fi novel *The Coming Race* [41] (see, e.g., [131, p. 522], [31, p. 194], and [103, p. xxxvi footnote 26]). When *Erewhon* was first published in 1872 it appeared anonymously and was widely taken to be Bulwer-Lytton's sequel to *The Coming Race*. Upon the announcement two months later that the author was in fact Butler, sales dropped by 90 percent [110, p. 155,158–159]. The literary scholar Marc Redfield goes as far as to question whether Eliot was intentionally "plagiariz[ing] the plagiarist" in *Shadows* (Butler himself having been suspected by some of plagiarising *The Coming Race*) [245]. This idea might not be as fanciful as it first appears if one considers that "Eliot successfully transmuted ideas into the form and structure of her novels; it is seldom sufficiently emphasized that this transmutation is in itself a key to her 'art'" [164, p. 25], and that in *Impressions* "[s]tories and phrases are freely borrowed. The concepts of originality and authorship are being questioned" [103, p. xxxiii]. However, speculations along these lines take us far beyond what can be proven, so we delve no further into this issue here.

3.4 The Late 19th Century

In the final two decades of the nineteenth century we see continuing allusions to self-reproducing machines, although nothing as protracted and explicit as the works of Butler, Marshall and Eliot.

In 1891 the controversial American author Wilford Hall published an article in which he argued that divine creation was still required to explain the *origin* of life even if we accept Darwin's arguments [133]. With echoes of William Paley before him (Sect. 2.2), Hall uses human technology as an analogy in his argument; while he appears to accept the possibility of a self-reproducing machine, he draws the line at machines that could produce output *more complex* than themselves: "No inventor, for example, constructs an ingenious machine and then expects that machine to evolve other inventions even still more complex than itself" [133, p. 162]. The fallacy in his argument was demonstrated some 50 years later, when John von Neumann did *exactly* that (as we will discuss in Sect. 5.1.1).

Turning to work with wider impact, H. G. Wells' seminal 1898 sci-fi novel *The War of the Worlds* alludes to the evolution of a bio-mechanical hybrid species: the Martians in the novel "may be descended from beings not unlike ourselves, by a gradual development of brain and hands ... at the expense of the rest of the body" [308]. As Leo Henkin remarks in his work *Darwinism in the English Novel*, Wells' Martians "have become practically mere brains, wearing different mechanical bodies according to their needs" [138, p. 255]. This brings to mind Butler's earlier image of the likely future degeneracy of the human body as we rely ever more upon machines in our daily activities (Sect. 3.1). These themes were later developed more extensively in the early twentieth century, most notably by J. D. Bernal, whose work we discuss in Sect. 4.2.1.

*   *   *

As demonstrated by the work described in this chapter, by the end of the nineteenth century the second major step in thinking about self-replicators—the birth of the idea of *evo-replicators* that can not only reproduce but also evolve—had already been accomplished. In the decades that followed, further explorations of the idea began to appear in short stories and plays aimed at a much wider general audience. At the same time, the increasingly common discussion of the idea began to catalyse a deeper exploration by scientists of the long-term implications of self-replicator technology. These developments are the subject of the following chapter.

Chapter 4
Robot Evolution and the Fate of Humanity: Pop Culture and Futurology in the Early 20th Century

By the turn of the twentieth century, the pace of technological development had created a more pressing need for considering where such progress might ultimately lead us. During this period, the exploration of potential futures of humanity in a world shared with self-reproducing, evolving machines was attracting a wider audience. Where Butler, Marshall and Eliot had led in considering these ideas in the late nineteenth century, others soon followed. In this chapter we discuss novels, sci-fi and other literature that explored self-reproducing machines in the early twentieth century, and we also cover speculative scientific work from this period. These works demonstrate the wider discussion of such ideas across society, and show that the current popularity of debates about advanced AI and AGI (e.g. [175, 134, 29, 284]) is actually a continuation of a public conversation that has been in progress for over a hundred years.

4.1 Literary Work

The growing popularity of the dystopian genre in early twentieth century literature was fuelled in part by a fear of how technology might negatively influence the development of human society [24]. Here we highlight works from the genre that involved ideas of machine self-reproduction and evolution.

4.1.1 E. M. Forster: The Machine Stops (1909)

E. M. Forster's short story *The Machine Stops* [115] was his only work of science fiction. It is now regarded as a classic of dystopian literature [107, p. 50].

The story pictures a future in which humans live underground in personal accommodation where corporeal needs are entirely satisfied by technology (the global, all-nurturing "Machine"). This leaves them free to concentrate on intellectual devel-

© Springer Nature Switzerland AG 2020
T. Taylor, A. Dorin, *Rise of the Self-Replicators*,
https://doi.org/10.1007/978-3-030-48234-3_4

opment, although it also renders them physically degenerate. Forster describes the Machine's "mending apparatus" that fixes problems and performs self-repair functions, evoking an early image of a machine capable of self-maintenance. It is the collapse of this functionality, brought about by the mending apparatus itself falling into disrepair, that brings the story to an apocalyptic end. Forster refers in passing to the Machine evolving new "food-tubes", "medicine-tubes", "music-tubes" and even "nerve-centres", but these ideas are not explored in detail.

As mentioned earlier (footnote 9, p. 22), Forster acknowledged the influence of Samuel Butler in his work—the vision in *The Machine Stops* of a future where an increasing dependency upon machines leads to the degeneracy of the human body certainly echoes some of Butler's concerns (Sect. 3.1). Forster's image of self-maintaining machines sustaining human life was further developed 20 years later by John Desmond Bernal, whom we discuss in Sect. 4.2.1.

4.1.2 *Karel Čapek:* R.U.R.: Rossum's Universal Robots *(1920)*

Themes of machine (collective) self-reproduction are present in Karel Čapek's play *R.U.R.: Rossum's Universal Robots*, published in 1920 and first performed in 1921 [56]. The robots[1] in the play were constructed from biochemical components and designed to resemble humans, but lacked "superfluous" capacities such as feelings or the capacity to reproduce. They were mass-produced in a factory to replace human workers with a cheaper, more productive alternative. The majority of the factory production was carried out by robots themselves, with only the most senior positions filled by humans. However, the complex formula for manufacturing the key "living material" was a closely-guarded secret, recorded by the factory's founder (Rossum) before his death and kept in a safe to prevent it from falling into the hands of competitors or the robots themselves.

One of the scientists in the factory experiments in making robots with more human-like feelings such as pain and irritability, but this leads to unintended and ultimately disastrous consequences when the robots come to despise their human masters and rise up against them. This, coupled with an unexplained crash in the human birth rate, leads to a stand-off where the robots surround the factory and the people within it, who are now the only surviving humans in the world. The humans realise that their only bargaining chip is the document that explains Rossum's formula, without which the robots would be unable to produce more of themselves and would therefore die out as a race as members of the current population fall into disrepair.

The climax of the play thus revolves around a struggle for the ownership of the written instructions that would allow the robots to collectively produce more of themselves—a struggle for the ownership of the robot's DNA, as it were.[2] This idea

---

[1] The play introduced the word "robot" into the English language.

[2] As it turns out in the play, this bargaining strategy of the humans comes to nothing when they realise that Rossum's document has already been destroyed. The play ends when the last remaining

of the collective reproduction of a society of robots reflects some of Butler's earlier ideas expressed in *Erewhon* (Sect. 3.1).

4.1.3 Early American Science Fiction (1920s–1950s)

The appearance of American pulp science fiction magazines in the 1920s, and their growing popularity over the decades that followed, provided a medium in which many writers explored the idea of self-reproducing robots and evolving machines.[3]

Perhaps the first example in this genre was the British writer S. Fowler Wright's story *Automata*, published in the American magazine *Weird Tales* in 1929 [313] (see Fig. 4.1). With echoes of Butler (Sect. 3.1), the story extrapolates the observed accelerating pace of technological development of the time into the far future, to a point when machines no longer rely on humans to service them. The machines become not only self-reproducing, but also able to *design* their own offspring. This ultimately leads to a complete takeover by machines and the extinction of the human race. The story views the takeover by machines as the inevitable next stage of evolution: similar to Butler's work, it suggests that the only way this could have been avoided was by "a war sufficiently disastrous to destroy the world's machinery and the conditions which could produce it" [313, p. 343]. As with Butler's and Eliot's work before it, the plot of *Automata* sounds a warning of the unpredictable long-term consequences of machine evolution:

> Even in the early days of the Twentieth Century man had stood in silent adoration around the machines that had self-produced a newspaper or a needle ... And at that time they could no more have conceived what was to follow than the first ape that drew the sheltering branches together could foresee the dim magnificence of a cathedral dome.
>
> S. Fowler Wright, *Automata*, 1929 [313, p. 344]

Three years later, in 1932, the influential American sci-fi writer and editor John W. Campbell published *The Last Evolution* [54], which also anticipated the eventual replacement of the human race by self-reproducing and self-designing machines.[4] However, Campbell's story is more optimistic than Wright's, foreseeing a period where humans live in peaceful and co-operative coexistence with intelligent machines, with human creativity complementing machine logic and infallibility. The end of the human race comes not at the hands of the intelligent machines, but when a species from another solar system invades Earth. The invasion prompts the machines to design a new super-intelligent machine to thwart the attack, and this itself

---

human, Alquist, meets two robots who seem to have developed the capacity for love and human-like reproduction, thereby giving hope that although the human race is about to die out, human-like life will continue.

[3] Some of the stories mentioned in this section are reprinted in [9]. Scans of most of the original publications are also available (for the purposes of private study, scholarship and research) from the *Luminist Archives* website at http://www.luminist.org/archives/SF/.

[4] As a student at MIT, Campbell had been taught by Norbert Wiener [139, p. 336], whose interest in self-reproduction within the field of cybernetics we describe in Sect. 5.5.

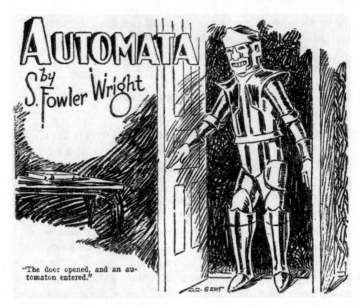

Fig. 4.1 Image from front page of S. Fowler Wright's sci-fi story *Automata* [313], published in 1929.

spawns further rounds of creation of more sophisticated machines—the final instantiation of which succeeds in repelling the invaders but is ultimately the only surviving species on Earth. Earlier in the story, the last two surviving humans console themselves while contemplating their fate:

> I think ... that this is the end ... of man ... But not the end of evolution. The children of men still live—the machines will go on. Not of man's flesh, but of a better flesh, a flesh that knows no sickness, and no decay, a flesh that spends no thousands of years in advancing a step in its full evolution, but overnight leaps ahead to new heights.
>   John W. Campbell, *The Last Evolution*, 1932 [54, p. 419]

Campbell's vision of a complementary coexistence of humans and intelligent machines is replaced by a darker image in his 1935 story *The Machine* (written under the pseudonym Don A. Stuart) [55]. In the story a human-like race on a planet named Dwranl[5] design a thinking machine that is set the task of making better versions of itself. The outcome is a machine that takes care of all of the race's basic needs. However, this ultimately leads to the degeneration of the race's intelligence, civility and its ability to look after itself—a similar fate to those described by Butler in *Erewhon* (Sect. 3.1), by Eliot in *Impressions* (Sect. 3.3) and by Forster in *The Machine Stops* (Sect. 4.1.1). The machine decides that its presence has become detrimental to the planet's inhabitants, for they are not engaging with it appropriately, but instead treating it like a god; it therefore resolves to leave the planet so that they can learn to live independently once more.

---

[5] The name being a near-anagram of *Darwin*.

Laurence Manning's *The Call of the Mech-Men* (1933) [197] also mirrors ideas first aired by Butler over 60 years earlier. Two explorers discover a group of extraterrestrial robots who have been living in underground caverns on Earth since their spaceship was damaged many tens of thousands of years earlier. The robots are amused when they hear of humankind's view of itself as master of its technology, remarking (in their stilted English): "Machine gets fed and tended under that belief! Human even builds new machines and improves year by year. Machines evolving with humans doing all work!" [197, p. 381].

Recurring themes of machine evolution and self-reproduction are seen in various stories over the following years. In Robert Moore Williams' *Robots Return* (1938), three robots from a faraway planet travel to Earth in search of information about the origins of their ancestors many thousands of years earlier [312]. To their surprise, they discover that they were originally designed by humans, and had been sent into space to accompany their creators in escaping a dying Earth. The humans did not survive the mission, but the robots did, settling upon a distant world; there, they reproduced and ultimately evolved into their current state.[6] Another tale of robots outliving their designers is Joseph E. Kelleam's *Rust* (1939), set on a post-apocalyptic Earth where human-designed robots have survived after humankind has been wiped out [160]. The robots try to design and build more of their kind before they succumb to erosion, but ultimately fail in their attempts. One further example is A. E. van Vogt's *M 33 in Andromeda* (1943), in which a spaceship of human explorers overcome an extraterrestrial intelligence the size of a galaxy by constructing a self-reproducing torpedo-manufacturing machine [299].

The most explicit exploration of machine self-reproduction and evolution in early science fiction is Philip K. Dick's *Second Variety* (1953) [85]. The story is set on Earth at the end of a long-running war between East and West, in which Western forces are driven to design killer robots to turn the tide on the battlefield. The robots are highly autonomous, with each generation gaining more sophisticated powers including self-repair and self-manufacture. The robots eventually become too dangerous for the human designers to be near, and are left to reproduce by themselves. Like the machines in Wright's *Automata* and Campbell's *The Last Evolution*, the robots in *Second Variety* eventually develop the ability to design their own offspring, and increasingly sophisticated and human-like species of killer robots begin to emerge. Echoes of these earlier stories are also seen when one of the human characters remarks "It makes me wonder if we're not seeing the beginning of a new species. *The new species. Evolution. The race to come after man*" [85].

Themes of machine self-repair, self-reproduction and evolution were central to various subsequent works by Dick. Another notable example is his 1955 short story *Autofac* [86], which ends with a vision of the seeds of self-reproducing manufacturing plants being launched into space.

Other sci-fi works from the 1950s featuring self-replicating machines include Robert Sheckley's 1955 short story *The Necessary Thing* [260], and Anatoly Dneprov's 1958 Russian work *Kraby Idut po Ostrovu* [89] (later published in English

---

[6] Lester del Ray later wrote a prequel to *Robots Return*, called *Though Dreamers Die* [80].

Fig. 4.2 Images taken from the original Russian version of Anatoly Dneprov's *Kraby Idut po Ostrovu* (*Crabs on the Island*) [89], published in 1958. From left to right, the images show three successive stages of the plot, from the inventor's arrival on the island with colleagues and their release of a single manufactured self-reproducing crab (along with "food" of various types of metal to be distributed around the island), to an evolutionary arms race developing among the growing crab population, to the eventual murder of the inventor by a large evolved crab that has noticed the valuable metal in his false teeth.

in 1968 as *Crabs on the Island* [90]). Dneprov's story has echoes of Dick's *Second Variety*, featuring small self-replicating robots designed as weapons. The robots are set loose on a desert island to compete against each other in an evolutionary arms race to produce ever more effective weapons. The experiment works, but not in the way that the machine's inventor had envisaged—he is eventually killed by one of the evolved machines (see Fig. 4.2).

Sources of further commentary on themes of robots and computers in early science fiction include [306] and [182]. Further substantial development of these themes was made during the 1960s—we provide references to some of the most distinguished of these later works in Chap. 6.

4.2 Scientific Speculation in the Early 1900s

Beyond these works of literature, the early twentieth century also saw continued speculations from the academic community on the long-term implications of machine self-reproduction and evolution. Here we look at the most notable example of this kind, written by the eminent British scientist J. D. Bernal.[7]

---

[7] We note in passing that the idea of self-reproducing machines was also touched upon—in a different context—by the respected American biogerontologist Raymond Pearl in his 1922 monograph *The Biology of Death* [232]. Emphasising that natural selection "makes each part [of an organism]

4.2.1 *J. D. Bernal:* **The World, The Flesh and the Devil** *(1929)*

John Desmond Bernal (1901–1971) conducted pioneering work on structural crystallography, and supervised the PhDs of two future Nobel laureates (Max Perutz and Dorothy Hodgkin) [145, 37]. Later in his career he also became interested in the origins of life [25]. In addition to his experimental science, he authored many works on history, science and society.

Bernal's first monograph—*"The World, the Flesh and the Devil: An Enquiry into the Future of the Three Enemies of the Rational Soul"* [26]—was perhaps his most futuristic writing.[8] In it, he discusses how one might examine the far future of humanity in a scientifically defensible way. Bernal begins by sign-posting the methodological and intellectual dangers to be avoided in such an endeavour, and discussing the unavoidable limitations. Keeping these issues in mind, he proceeds to explore what might be said of the three major kinds of struggle facing humanity: against the forces of nature and the laws of physics in general ("the world"); against biological factors including ecology, food, health and disease ("the flesh"); and against psychological factors including desires and fears ("the devil").

Writing before the advent of space travel, atomic energy or computers, Bernal first tackles how humankind might overcome the challenges that arise from the material world. He argues that limitations of land and energy in the world will eventually compel us to colonise space: "On earth, even if we should use all the solar energy which we received, we should still be wasting all but one two-billionths of the energy that the sun gives out. Consequently, when we have learnt to live on this solar energy and also to emancipate ourselves from the earth's surface, the possibilities of the spread of humanity will be multiplied accordingly" [26, p. 22]. After discussing plausible technologies for powering a spaceship to escape the Earth's gravitational field and then to travel in space, he goes on to imagine how humans might set up permanent space colonies.

Bernal proposes a "spherical shell ten miles or so in diameter" [26, p. 23] which could provide a habitable environment for twenty or thirty thousand inhabitants. After discussing how the construction of a sphere might be bootstrapped from a basic design built largely of materials mined from an asteroid, Bernal continues with a description of the organisation of a mature sphere. It is imagined as "an enormously complicated single-celled plant" [26, p. 23] with a protective "epidermis", complete with regenerative mechanisms to protect against meteorites, mechanisms

---

just good enough to get by", he continues: "The workmanship of evolution, from a mechanical point of view, is extraordinarily like that of the average automobile repair man. If evolution happens to be furnished by variation with fine materials, as in the case of the nervous system, it has no objection to using them, but it is equally ready to use the shoddiest of endoderm provided it will hold together just long enough to get the machine by the reproductive period" [232, p. 148]. This being the case, it is "conceivable that an omnipotent person could have made a much better machine, as a whole, than the human body which evolution has produced, assuming, of course, that he had first learned the trick of making self-regulating and self-reproducing machines, such as living machines are" [232, p. 148].

[8] Arthur C. Clarke later described it as "the most brilliant attempt at scientific prediction ever made" [63, p. 410].

for the capture of meteoric matter to be used as raw material for the growth and propulsion of the sphere, systems for energy production from solar energy, stores for basic goods such as solid oxygen, ice and hydrocarbons, and mechanisms for the production and distribution of food and mechanical energy. The sphere would also have mechanisms for recycling all waste matters, "for it must be remembered that the globe takes the place of the whole earth and not of any part of it, and in the earth nothing can afford to be permanently wasted" [26, p. 25].

The inhabitants of these globes in space would not be isolated, but would be in wireless communication with other globes and with the Earth. In addition, there would be a constant interchange of people between the globes and the Earth via interplanetary transport vessels. Having set out how the globes might function to sustain life as "mini-earths", Bernal imagines a yet more ambitious scenario:

> However, the essential positive activity of the globe or colony would be in the development, growth and reproduction of the globe. A globe which was merely a satisfactory way of continuing life indefinitely would barely be more than a reproduction of terrestrial conditions in a more restricted sphere.
>
> J. D. Bernal, *The World, The Flesh and the Devil*, 1929 [26, p. 27]

Hence, the globe is conceived of as a fully self-maintaining and self-reproducing unit akin to a living organism.[9] Bernal discusses various ways in which a globe might construct another globe:

> ... either by the crustacean-like development in which a new and better globe could be put together inside the larger one, which could be subsequently broken open and re-absorbed; or, as in the molluscs, by the building out of new sections in a spiral form; or, more probably, by keeping the even simpler form of behavior of the protozoa by the building of a new globe outside the original globe, but in contact with it until it should be in a position to set up an independent existence.
>
> J. D. Bernal, *The World, The Flesh and the Devil*, 1929 [26, p. 27]

Once the globes are equipped with the capacity for self-reproduction, Bernal further envisages how an evolutionary pressure to explore might arise among a population of globes:

> As the globes multiplied they would undoubtedly develop very differently according to their construction and to the tendencies of their colonists, and at the same time they would compete increasingly both for the sunlight which kept them alive and for the asteroidal and meteoric matter which enabled them to grow. Sooner or later this pressure ... would force some more adventurous colony to set out beyond the bounds of the solar system.
>
> J. D. Bernal, *The World, The Flesh and the Devil*, 1929 [26, p. 29]

The enormous challenges that would be faced in travelling interstellar distances are addressed, but Bernal argues that such a vision is nevertheless reasonable to consider: "once acclimatized to space living, it is unlikely that man will stop until he has roamed over and colonized most of the sidereal universe, or that even this will be the end. Man will not ultimately be content to be parasitic on the stars but will invade them and organize them for his own purposes" [26, p. 30].

---

[9] To apply a modern term first introduced by the biologists Humberto Maturana and Francisco Varela to describe living systems, we might describe the globe as an *autopoietic* organisation [199].

Fig. 4.3 Concept art depicting the construction of a Bernal Sphere by Don Davis, prepared for the 1975 NASA Summer Study on Space Settlements [154]. This image is from a collection of art associated with the study available at https://settlement.arc.nasa.gov/70sArt/art.html.

Moving next to the possibilities of how our own bodies might develop in the distant future, Bernal imagines that humankind will increasingly replace and augment body parts with synthetic alternatives—a vision previously explored by Butler (Sect. 3.1) and H. G. Wells (Sect. 3.4) among others.

Turning to the activities such advanced beings might pursue, Bernal suggests that, among other important scientific questions,[10] there would surely be intensive further study of life processes, and the creation of synthetic life. However, "the mere making of life would only be important if we intended to allow it to evolve of itself anew ... [however] artificial life would undoubtedly be used as ancillary to human activity and not allowed to evolve freely except for experimental purposes" [26, p. 45].[11]

Bernal's vision of the relationship between the future evolution of humans and machines is more symbiotic than the futures imagined by Forster and Čapek: "Normal man is an evolutionary dead end; mechanical man, apparently a break in organic evolution, is actually more in the true tradition of a further evolution" [26, p. 42]. This perspective is more in line with the ideas expressed by Butler in *Lucubratio Ebria* (Sect. 3.1), and with those of sci-fi authors such as John W. Campbell (Sect. 4.1.3). Bernal suggests that the main barriers towards progress in these areas would arise from human psychology—in addition to having the desire for progress, we must also "overcome the quite real distaste and hatred which mechanization has already brought into being" [26, p. 55].[12] Various ways in which such barriers may be overcome are suggested, but Bernal does not discount the alternative possibility that we ultimately find ways of living a simpler yet more satisfying life that is not occupied by science or art but more at one with nature.[13] He also considers a third possibility—"the most unexpected, but not necessarily the most improbable" [26, p. 56]—that the future evolution of humanity might diverge, with one race following the natural path and another race following the intellectual and technological path.

Nearly fifty years after the publication of *The World, The Flesh and the Devil*, Bernal's idea of a space globe inspired one of the concepts for human space colonies

---

[10] However, Bernal did not see the future intellectual development of humanity resting on science alone: "just as all the branches of science itself are coalescing into a unified world picture, so the human activities of art and attitudes of religion may be fused into one whole action-reaction pattern of man and reality" [26, p. 54] (a similar idea was later explored by the philosopher John Macmurray [195, ch. 10]).

[11] In this section of the text, Bernal refers to an essay published in 1927 by the Scottish physicist L. L. Whyte, entitled *Archimedes, or The Future of Physics* [309]. In the context of examining the convergence of the sciences of physics, biology and psychology in the study of life, Whyte discusses approaches to creating synthetic life using chemical components. He argues that higher-level intelligence would have to be evolved rather than designed, and estimates minimum times that might be required to achieve various evolutionary outcomes (ranging up to one million years for mammalian-level intelligence). Whyte—regarded as a maverick by some [310, p. ix]—proposes the establishment of an *International Institute for Evolutionary Research* to oversee such a massively long-term synthetic evolutionary study of living systems [309, pp. 47–65].

[12] Such feelings, Bernal says, are a natural result of the fact that "[t]he human mind has evolved always in the company of the human body" [26, p. 60].

[13] In contrast to Butler (Sect. 3.1), who suggested that mankind was already past the point of no return in technology to allow such a reversal.

developed during a 1975 NASA Summer Study on Space Settlements (see Fig. 4.3). The concept was named the "Bernal sphere" in honor of his work [154, p. 48].

4.2.2 The Widening Impact of Ideas

The work described in this chapter, both by sci-fi authors and by scientists, brought the idea of machine self-reproduction and evolution to a wider audience. While scientists paid closer attention to theory and to practical details, that is not to say that the ideas set out in the literary works covered above were not taken seriously. In particular, Čapek's play *R.U.R.* became internationally well known and influential within a few years of its release [57, p. 9]—Winston Churchill, for example, referred to it in his 1931 essay *Fifty Years Hence*, in which he discussed what could be said of how the world might develop in the coming decades [61].[14]

Another sci-fi author from the period whose work attracted widespread attention was Olaf Stapledon. His 1937 novel *Star Maker* [270] featured an extensive further exploration of the possibilities for humanity's descendants living on self-sustaining artificial planets, as originally proposed by Bernal.[15] Stapledon's work was admired by figures as diverse as Winston Churchill, Arthur C. Clarke, Freeman Dyson and Virginia Woolf.[16]

*   *   *

As we have seen, the early decades of the twentieth century saw an increasingly wide-spread discussion of evo-replicator technology in fictional literature, accompanied by a pioneering, more scientifically-grounded discussion by J. D. Bernal of its potential implications for the far future of humanity. This set the stage for the work of the mid-twentieth century we are about to discuss in the next chapter, which saw a rapid proliferation of contributions from scientists. Along with more detailed theoretical discussions of the potential uses of this technology, the mid-twentieth century also saw the first rigorous work on a design theory for self-reproducing machines,

---

[14] Churchill's essay also echoed some of the themes raised by Butler (Sect. 3.1), such as the pace of technological development having progressed so far that "Mankind has gone too far to go back, and is moving too fast to stop" [61]. Churchill recognised, as did Bernal before him, that the human mind might struggle to keep up with the rate of development of technology. But in contrast to Bernal (whose work looked over far longer horizons), Churchill thought the solution was not for us to overcome our distaste of mechanization, but to nurture "an equal growth in Mercy, Pity, Peace and Love, [without which] Science herself may destroy all that makes human life majestic and tolerable" [61].

[15] However, Stapledon did not explore the possibilities of self-reproduction of the globes, as Bernal had done.

[16] For details, see Gregory Benford's foreword in the SF Masterworks edition of Stapledon's *Last and First Men* [271], and the quotes at the start of the SF Masterworks edition of *Star Maker* [272].

and the first implementations of self-replicator systems in software and in hardware. This period therefore marks the attainment of the third and final step in the history thus far of the intellectual development of thought about self-replicators (Sect. 1.6). It is to these groundbreaking developments that we now turn.

Chapter 5
From Idea to Reality: Designing and Building Self-Reproducing Machines in the Mid-20th Century

Alan Turing's development of a theory of universal computation in the 1930s [293], followed by the appearance of the first digital computers in the 1940s, allowed people to experiment with processes of *logical* self-reproduction—that is, self-reproduction implemented in software without the extra difficulties entailed by physical self-reproduction. The 1940s and, in particular, the 1950s saw the emergence of the first rigorous theoretical work on the design of self-reproducing machines, and of the first implementations of artificial self-reproducing systems in software (logical self-reproduction) and in hardware (physical self-reproduction).

We have now reached the point where some (but certainly not all) of this work has been widely covered in other publications.[1] In this chapter, we review the most significant developments during this period. We begin by looking at John von Neumann's contributions to the theory of the subject (Sect. 5.1.1); although this is covered in other reviews, we include it because of its significance, because some other reviews do not emphasise von Neumann's interest in the *evolutionary* potential of his self-reproducing machines,[2] and so that we can present it in the context of what came before and afterwards. We also highlight other work that has not been so widely discussed, such as the pioneering experiments in software evolution by Nils Aall Barricelli (Sect. 5.2.1), Konrad Zuse's early thoughts about the design and application of self-reproducing machines (Sect. 5.4.2) and Andrei Kolmogorov's lectures and writing on the subject (Sect. 5.5). Furthermore, while the work of others discussed in this section—such as Lionel Penrose (Sect. 5.3.1) and Homer Jacobson (Sect. 5.3.2)—*has* been reported elsewhere, we provide some additional technical and biographical details that have not been highlighted in other reviews.

As we mentioned at the start of the book (Sect. 1.3), the work of this period sees the emergence of a third flavour of self-replicator; in addition to continued interest in standard-replicators and evo-replicators, we observe the introduction of the idea that a self-reproducing machine could be used as a general-purpose manufacturing

---

[1] In Chap. 6 we provide references to the most significant of these other reviews.

[2] See [203] for an insightful discussion.

machine—a *maker-replicator*. This concept was first seen in von Neumann's work as part of his design theory for self-replicators.

5.1 Theory of Logical Self-Reproduction

The initial development of a theoretical basis for the design of self-reproducing machines in this period was due almost single-handedly to the Hungarian-American scientist John von Neumann in the late 1940s and early 1950s. In the later 1950s and early 1960s there were also contributions from the field of cybernetics, which we summarise in Sect. 5.5. However, these were far less substantial and influential than von Neumann's work.

5.1.1 John von Neumann (1948)

By the late 1940s, von Neumann (1903–1957) had become interested in developing general principles for the design of immensely complex machines that could tackle pressing scientific and engineering problems. Looking for inspiration in biology, he developed a particular interest in the capacity of biological organisms for self-reproduction, and in the observed evolutionary increase in complexity of some organisms over time. He saw self-reproduction and evolution as a means to an end—to produce complex automata that could solve real problems—rather than an end in itself [303, p. 92].

Noting that some processes (e.g. crystal growth) display a trivial kind of self-reproductive ability, von Neumann clarified that the kind of self-reproduction process of interest in his work must have "the ability to undergo inheritable mutations as well as the ability to make another organism like the original" [303, p. 86]. In other words, an important aspect of his research was the development of design principles for evo-replicators. But his focus was more specific than the design of evo-replicators in general; in addition to being able to evolve, his self-reproducing machines would need the ability to perform other arbitrary operations, and the complexity of these operations should be able to increase each time the machine reproduced, to eventually generate machines capable of solving difficult problems.

By 1948 von Neumann had already proposed a general abstract architecture that might be used to address the problem [302]. His solution was inspired by Alan Turing's work on universal computing machines [293],[3] but he modified Turing's design such that the input and output operations acted upon structures composed of the same materials out of which the machine itself was composed [303, pp. 75–76]. Von Neumann's machine could therefore construct other machines as part of its operation.

---

[3] Subsequent authors have also discussed the connection between von Neumann's design and Kleene's Recursion Theorem (see, e.g., [59, pp. 28–29], [157, pp. 367–371]).

Von Neumann discussed the level at which the individual parts of the architecture should be defined:

> by choosing the parts too large, by attributing too many and too complex functions to them, you lose the problem at the moment of defining it. ... One also loses the problem by defining the parts too small ... In this case one would probably get completely bogged down in questions which, while very important and interesting, are entirely anterior to our problem. ... So, it is clear that one has to use some common sense criteria about choosing the parts neither too large nor too small.
>
> John von Neumann, 1949 [303, p. 76]

He envisaged a universal construction machine comprising three subcomponents: a *building* unit, a *copying* unit and a *control* unit. The building unit, when fed an information tape bearing an encoded description of the design of an arbitrary machine, would build that machine from the description; the copying unit, when fed an information tape, would produce a second copy of the tape; the control unit would coordinate the actions of the other two units.

Von Neumann showed that if this universal construction machine was fed an encoded description of *itself*, the result was a self-reproducing machine (see Fig. 5.1). The fundamental aspect of the design, which circumvented a potential infinite regress of description, was the dual use of the information tape—to be interpreted as instructions for creating a duplicate machine (by the building unit) and to be copied uninterpreted for use in the duplicate (by the copying unit).[4]

In addition to self-reproduction, the design also satisfied von Neumann's requirement that the machines should be able to perform other tasks of arbitrary complexity. He showed how such machines could produce offspring capable of performing more complex tasks than their parents, and how such increases in complexity could come about through heritable mutations to the information tape. Hence, these kinds of machines have the potential to participate in an evolutionary process leading to progressively more complex machines which could perform arbitrary tasks in addition to self-reproduction.

To summarise, it was von Neumann's desire to create self-reproducing machines that could perform arbitrary useful tasks in addition to reproduction (where the complexity of these tasks could increase over the course of evolution), coupled with his inspiration from Turing's idea of a universal computing machine, which drove him to the idea of a self-reproducing universal construction machine—a *maker-replicator* (and indeed an evolvable maker-replicator, or *evo-maker-replicator*).

Having established a general theory of the logical design of an evo-maker-replicator, von Neumann planned to explore the topic experimentally with the aid of a series of models [10, pp. 374–381].[5] The first of these was the so-called "cellular

---

[4] Von Neumann's description of the logical design of a self-reproducing machine can equally be applied to the reproductive apparatus of biological cells. However, although his work pre-dated the unravelling of the details of DNA replication by some years, it had little impact on developments in genetics and molecular biology [33, pp. 32–36].

[5] Arthur W. Burks, who edited von Neumann's posthumously published book *Theory of Self-Reproducing Automata*, commented that von Neumann "intended to disregard the fuel and energy problem in his first design attempt. He planned to consider it later" [303, p. 82].

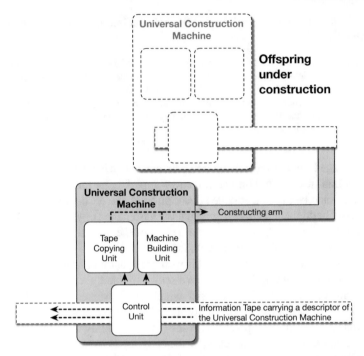

Fig. 5.1 Schematic of von Neumann's design for a universal construction machine capable of self-reproduction and evolution. The machine in the lower part of the image has been supplied with a description of its own design on its information tape. It is shown in the process of using this information to construct a copy of itself and of the information tape, as shown in the upper part of the image.

model" based upon a simple two-dimensional grid of squares (cells) in which, at any given time, each square could be in one of a small number of possible states.[6] The other models had been planned to progressively move away from discrete models to ones with more "lifelike" properties such as continuous dynamics and probabilistic operation. Von Neumann emphasised that the design of the machines would depend critically upon the nature of their environment: "it's meaningless to say that an automaton is good or bad, fast or slow, reliable or unreliable, without telling in what milieu it operates" [303, p. 72]. Studying self-reproduction in a series of progressively more complex environments was therefore a sensible approach.

Before his early death in 1957, von Neumann was only able to produce a detailed design for the first of these, the cellular model. Despite the ingenuity of the design,

---

[6] The cellular model was based upon a Cellular Automaton (CA) design, which von Neumann had conceived with his colleague Stanislaw Ulam, and is the first work to use the now-popular CA formalism. Von Neumann had originally conceived of a more complex "kinematic model", but Ulam observed that a discrete model would be more analytically tractable [10, p. 375]. In Ulam's biography, he himself recalls informal discussions that he had with Stanislaw Mazur of the Lwów Polytechnic Institute in 1929 or 1930 on "the question of the existence of automata which would be able to replicate themselves, given a supply of some inert material" [297, p. 32].

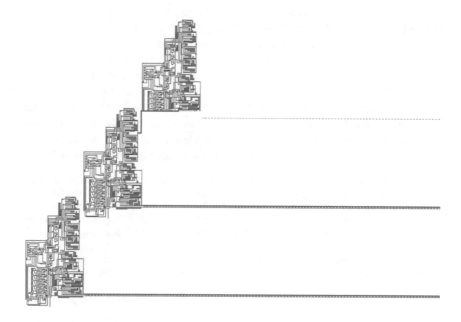

Fig. 5.2 Image from a recent full implementation of a self-reproducing automata based upon von Neumann's design. This version, from 2008, employs Renato Nobili and Umberto Pesavento's universal constructor design with tape design by Tim Hutton, and is implemented in the Golly cellular automata software (http://golly.sourceforge.net/).

to actually implement it on a computer was far beyond the available computational power of the time. Just a single machine and information tape bearing an encoded description of the machine would require approximately 200,000 cells [161, p. 66]; to simulate the production of just one offspring would require at least double that number. Indeed, the first implementation of the cellular model (with some simplifications) did not appear for another forty years [238], followed by several more recent versions—an example is shown in Fig. 5.2.

Von Neumann's ideas soon spread beyond the scientific community, and by the mid-1950s they were the topic of articles in popular science magazines (e.g. [161][7]). However, despite its very substantial achievements, the work only partially addressed the problems involved in designing *physical* self-reproducing machines. In particular, the work did not address issues regarding fuel and energy, nor did the architecture include any kind of self-maintenance system—the basic machine as presented would be very susceptible to any kind of external damage from the environment. We return to these issues of extending von Neumann's design to handle real-world self-reproduction in Sect. 7.3.1.

---

[7] By arrangement with von Neumann [303, p. 95], Kemeny's article [161] was based upon lectures that von Neumann had delivered at Princeton University in 1953 and upon parts of the manuscript for [303].

The universal construction machine is like a massively complex Lego kit where the instruction manual is also built from Lego, and where the kit can read, execute and copy the manual itself. How such a kit might originally come about was not part of the problem addressed by von Neumann. However, *if* such a kit *did* exist, then it could operate and evolve purely on a diet of elementary Lego bricks—the complexity of its operation is due mostly to the organisation of the machine and to the information contained in the instruction manual, rather than the individual bricks. We will return to the question of *origins*—of how complex self-reproducing systems might arise in an environment in the absence of an original designer—in Sect. 7.3. Von Neumann restricted himself to the following remarks on the topic:

> ... living organisms are very complicated aggregations of elementary parts, and by any reasonable theory of probability or thermodynamics highly improbable. That they should occur in the world at all is a miracle of the first magnitude; the only thing which removes, or mitigates, this miracle is that they reproduce themselves. Therefore, if by any peculiar accident there should ever be one of them, from there on the rules of probability do not apply, and there will be many of them, at least if the milieu is reasonable. But a reasonable milieu is already a thermodynamically much less improbable thing. So, the operations of probability somehow leave a loophole at this point, and it is by the process of self-reproduction that they are pierced.
>
> John von Neumann, 1949 [303, p. 78]

Freeman Dyson, whose contributions to the subject we discuss in Sect. 6.3, summarised the significance of von Neumman's work as follows:

> Von Neumann believed that the possibility of a universal automaton was ultimately responsible for the possibility of indefinitely continued biological evolution. In evolving from simpler to more complex organisms you do not have to redesign the basic biochemical machinery as you go along. You have only to modify and extend the genetic instructions. Everything we have learned about evolution since 1948 tends to confirm that von Neumann was right.
>
> Freeman Dyson, 1979 [96, p. 196]

Von Neumann's contribution to the theory of self-reproducing machines has been hugely influential in later work on the topic. We provide an overview of these more recent developments in Chap. 6.

5.2 Realisations of Logical Self-Reproduction

Beyond von Neumann's development of a theoretical basis for the the design of evo-maker-replicators, the 1950s also saw the first work in actually *building* artificial self-reproducing machines, both in software and hardware. These projects took a very different approach to the subject, building standard-replicators or evo-replicators using simple elementary units that could combine to form self-reproducing configurations. In contrast to von Neumann, the authors of these projects were motivated (at least in part) by questions concerning the origin of life on Earth.

5.2.1 Nils Aall Barricelli (1953)

The story of Nils Aall Barricelli (1912–1993) is remarkable in many ways. To summarise what we will discuss below, in 1953 he became the first person to perform
experiments in logical self-reproduction and artificial evolution on computers. His
interest in achieving ongoing, *open-ended* evolution of progressively more complex
digital organisms, and his conviction that he was *instantiating*, rather than merely
simulating, evolutionary processes in a computational medium, place his research
firmly within the present-day scientific discipline of Artificial Life.[8] Yet, despite
his pioneering achievements, he is still relatively unknown even within the contemporary ALife community, where he might justifiably be regarded as one of the
founding fathers of the field.[9]

Born in Rome to an Italian father and Norwegian mother, Barricelli moved to
Norway in 1936 at the age of 24.[10] He took up a lecturing position in physics at
the University of Oslo, although he maintained a wide range of academic interests
throughout his life. By the late 1940s, he had become interested in the theory of
symbiogenesis, introduced in the early twentieth century by the Russian botanists
Konstantin Merezhkovsky and Boris Kozo-Polyansky [173, 174]. According to this
theory, "the genes [of a cell] ... spring from originally independent virus or virus-
like organisms" [16, p.74].

Having performed some initial simulation experiments of the theory by hand on
graph paper, Barricelli saw the potential for running greatly expanded experiments
on an electronic computer. He successfully applied to join von Neumann's group
at the Institute for Advanced Studies in Princeton as a visiting researcher, where he
would have access to the group's recently-built computer, the "IAS Machine." Upon
his arrival in January 1953, Barricelli worked night-shifts when the machine was
free from its daytime employment on national defence-related work. He conducted
a series of experiments in 1953 and during subsequent return visits in 1954 and 1956
[15, p. 88].

Barricelli was interested in studying the simplest possible systems that could
display evolutionary processes leading to "the formation of organs and properties
with a complexity comparable to those of living organisms" [15, p. 73]. He therefore
assumed that the individual elements in his systems were already equipped with
some capacity for self-reproduction. This was in sharp contrast to von Neumann's
work, which aimed to understand how that capacity might be built into a system

---

[8] As mentioned in Sect. 1.4, the study of open-ended evolutionary systems has attracted significant
renewed interest within the Artificial Life and AI communities within the last few years [281, 269,
226]. See Sect. 6.1 for further discussion of this topic.

[9] Furthermore, in 1963 he proposed early ideas for what is now known as DNA computing [16,
pp. 121–122]. In recent years, Barricelli's work has started to receive slightly more coverage in
the academic literature (e.g. [112, 121]), in books (e.g. [97, 98]) and in magazine articles (e.g.
[120, 130]).

[10] Sources of biographical details in this paragraph and the next include [98, pp. 225–228] and
[121, pp. 29–31].

composed of individual elements that were *not* themselves self-reproducing. We will discuss the difference between these two approaches in more detail in Sect. 7.3.

Barricelli's experiments were conducted in a virtual one-dimensional world—a strip of discrete square units updated in discrete time steps. Each square in the world could either be empty (represented by the value 0) or occupied by a small positive or negative non-zero integer.[11] The state of a square at the next time step was determined by its state at the current time step together with the state of its neighbours. The exact details of this mapping were determined by the system's *update rule*.

By exploring a series of different update rules, Barricelli observed the emergence of progressively more complex forms of behaviour.[12] His earliest hand-drawn simulation experiments involved update rules that implemented what he described as the bare necessities of a Darwinian evolution process, such that individual numbers in his one-dimensional world were (1) self-reproducing and (2) capable of undergoing heritable mutation [15, p. 71]. Specifically, if a square contained a non-zero integer n at time t, then at time $t+1$ the same integer n would also be copied into the square n places to the left or right of the original, depending on whether it was positive or negative. If two integers tried to reproduce into the same square, then a different integer was placed in the square, thereby introducing a mutation into the system (Barricelli experimented with various rules for deciding what the mutated integer should be).

Although the system was capable of reaching a stable state through a "process of adaptation to the environmental conditions" [15, p. 72], Barricelli was unsatisfied with the end result: "No matter how many mutations occur, the numbers ... will never become anything more complex than plain numbers" [15, p. 73]. It was apparent that something more was required beyond the basic Darwinian properties of reproduction and heritable mutation.

For Barricelli, the symbiogenesis theory potentially supplied the missing ingredient. He made changes to the update rules for his one-dimensional world such that each number could only reproduce with the support of a "helper" number in a specific nearby location. The helper would also dictate the location into which the number reproduced. In this way, reproduction in the system was no longer at the level of "plain numbers": if it was to happen in a sustained manner, it now required the mutual co-operation of spatially organised aggregates of numbers. Barricelli did indeed see such aggregates emerge, and called them "numerical symbioorganisms"

---

[11] The design was a kind of one-dimensional cellular automaton (see footnote 6, p. 44), considerably simpler than the two-dimensional CA employed in von Neumann's cellular model. In the experiments Barricelli conducted at IAS, the world was a cyclic array with 512 squares, where the state of each square could lie in the range -40 to +40 [285, p. II-83].

[12] Barricelli's first results were published, in Italian, in the journal *Methodos* in 1954 [12], followed in 1957 by an expanded English-language version in the same journal [13]. The most detailed presentation of his work appeared in two papers published in the theoretical biology journal *Acta Biotheoretica* in 1962–63 [15, 16], with a more condensed summary in *The Journal of Statistical Computation and Simulation* in 1972 [17].

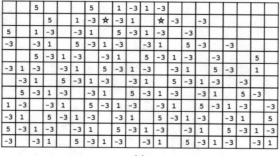

(a)

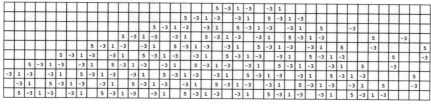

(b)

Fig. 5.3 Examples of (a) emergence and (b) self-reproduction of a symbioorganism in Barricelli's system described in [15]. Each row in these figures represents the state of the one-dimensional world at a given time. Each successive time-step of the system is plotted below the last, so the figure shows time advancing from the top to the bottom of the plot. The symbiogorganism that emerges in (a) is [5,-3,1,-3,_,-3,1]. The same symbioorganism is then used to seed the system in the top row of (b). The update rules used in these examples are as follows: (1) Each number n is moved to a position n squares to the right (if n is positive) or n squares to the left (if n is negative) in the next row; (2) If a collision occurs between two different numbers in the new row, the square remains empty—this is indicated by the star symbols in the second row of (a); (3) If a collision occurs between two identical numbers, that number is written once to the square only; (4) If the new number n lands in a square which has another number m directly above it in the preceding row, a second copy of n is produced m squares to the right (if m is positive) or to the left (if m is negative) of the position of the *original* n (this rule can be iterated if the second copy of n also lands in a square with another number directly above it).

[15, p. 80].[13] His subsequent experiments on the IAS Machine were devoted to studying their properties.

Among the general properties observed in the symbioorganisms,[14] Barricelli gave examples of the following: self-reproduction, crossing, great variability, mu-

---

[13] Symbioorganisms can be viewed as examples of what have now become known as collectively autocatalytic sets (see, e.g., [159]).

[14] Barricelli initially came up against the problem of the system reaching a state of "organized homogeneity" after a few hundred generations [15, p. 88]. This occurred when a single variety of symbioorganism had invaded the whole world and no further change was observed. He made several attempts to avoid this behaviour, and finally adopted a successful approach that involved two modifications to the original design. First, he divided the world into four separate areas and used slightly different update rules in each area, and second, he ran several different evolution experiments in parallel, exchanging the content of subsections of the world between two universes

Fig. 5.4 A sample output from a modern reimplementation of Barricelli's symbioorganism experiments written by Alexander Galloway. Each horizontal strip of the image represents the state of the two-dimensional world at a given time. Each successive time-step of the system is plotted below the last, so the figure shows time advancing from the top to the bottom of the plot. The different colours represent different symbioorganisms, and their rise and fall over time can be seen in the expanding and contracting patterns they create in the plot.

tation, spontaneous formation, parasitism, repairing mechanisms and evolution [15, pp. 81–88] (see Fig. 5.3).[15] In evolution experiments which ran for thousands of generations, successions of various dominant species of symbioorganism were seen

---

every 200 or 500 generations [15, p. 89]. The topic of diversity maintenance in computational evolutionary systems is still very much a matter of active research [70].

[15] Barricelli clarified his use of biological terminology at the start of the paper, justifying his use of such terms because they "were easy to remember and [they made] analogies with biological concepts immediately clear to the reader without requiring tedious explanations." However, he cautioned that "[t]he terms and concepts used in connection with symbioorganisms are in no case identical to biological concepts; they are mathematical concepts ..." [15, p. 70].

to emerge (see Fig. 5.4).[16] Barricelli also performed competition experiments where he pitted some of the successful species of symbioorganisms that evolved during the runs against each other. The results led him to conclude that "it is clear that the ability to survive is improved by evolution" [15, p. 92]. He also noted that the phenomenon of crossing of genetic material between different symbioorganisms appeared at an early stage in all of his experiments, and remarked that "[t]he common idea that living beings may have existed for a long time before crossing mechanisms appeared is hardly consistent with a symbiogenetic interpretation" [15, p. 93].

Having observed in his competition experiments that the symbioorganisms could exhibit evolutionary improvements, Barricelli wondered "whether it would be possible to select symbioorganisms able to perform a specific task assigned to them" [16, p. 100]. To investigate this question, he ran a further set of experiments on an IBM 704 computer[17] at the A. E. C. Computing Center at New York University in 1959 and 1960 [16, p. 107].

The new experiments involved a relatively small conceptual change—if two numbers tried to copy themselves into the same memory location, rather than causing a mutation as in the original experiments, the rival symbioorganisms would now compete in a game where the result would determine which one was allowed to occupy the contested location. The game chosen was *Tac Tix*,[18] and Barricelli devised a method for interpreting a specific consecutive sixteen number sub-sequence of a symbioorganism as a strategy for playing it [16, p. 103–106]. The symbioorganisms' new-found ability to perform other tasks as well as reproduce was one they shared with von Neumann's theoretical architecture—although Barricelli's design was much more basic, involving a fixed interpretation of the numbers as a strategy for a specific game, in contrast to von Neumann's interpretation machinery which was itself able to evolve and which formed a component of a universal construction machine.

After some experimentation, the 1960 experiments yielded several runs in which an evolutionary improvement was observed in the fraction of symbioorganisms playing the game at a non-zero skill level.[19] Although the results were not consistent across runs, and in at least one of the reported runs a parasite evolved which

---

[16] Although he reported that in the latter stages of the runs "evolution proceeded at a slower rate until generation 5000 when the experiment was discontinued" [15, p. 90].

[17] This allowed Barricelli to use worlds of size 3072 locations—almost a six-fold increase compared to the experiments on the IAS Machine.

[18] This game was described by Martin Gardner and based upon the popular game of Nim [122].

[19] Barricelli used "the number of correct final decisions of the winner" as the measure of skill level, and reports that human novice players typically show a skill-level of 1 or 2 in their first five games played [16, p. 107].

disrupted the further improvement of the host,[20] Barricelli summarised the significance of the results as follows:

> ... the value of the results presented does not primarily rest on the possibilities for practical applications, but on their biotheoretical significance. ... It has been shown that given a chance to act on a set of pawns or toy bricks of some sort the symbioorganisms will 'learn' how to operate them in a way which increases their chance for survival. This tendency to act on any thing which can have importance for survival is the key to the understanding of the formation of complex instruments and organs and the ultimate development of a whole body of somatic or non-genetic structures.
>
> Nils Aall Barricelli, 1963 [16, p. 117]

In addition to the preceding remarks about the fixed mechanism for interpreting a symbioorganism's game strategy, it has recently been argued that the approach of introducing a predefined game to be played restricted the open-ended evolutionary potential of the system by "flattening complex interactions back into linear causal chains" [121, p. 41]. In other words, the game was not co-evolving along with the organisms. Barricelli himself seems to have been sensitive to these criticisms, and in his final published work in the area in 1987 he sets out some suggested modifications to his original design whereby symbioorganisms have the opportunity to evolve operational elements such as membrane-like structures [18].[21] Such elements would allow the symbioorganisms to control their local environment and thereby improve their chances of survival and reproduction in a way that operated within the general laws of the world instead of via a special-case predefined game to be played. He discussed how a more pronounced genotype–phenotype distinction would emerge in such a system, accompanied by what could be regarded as a "genetic language" to specify a symbioorganism's operational elements and their organisation [18, pp. 143–144]. At the age of 75, Barricelli was clearly not intending to commence a new line of experimental work himself, and the paper was written as a set of suggestions for a future programmer to follow.[22] He died six years later in 1993, at the age of 81.

---

[20] See [16, pp. 114–116] for details. Barricelli summarises that "the parasite never developed an independent game strategy to any degree of efficiency and was entirely dependent on its host organism for game competitions with, and transmission of the infection to, uninfected hosts [16, p. 125]. In one line of subsequent research, he worked with colleagues on a quite different evolutionary system more resembling a genetic algorithm, and obtained better results in evolving strategies for poker [246], [17, pp. 119–122].

[21] He had already discussed the importance of membrane structures in the early evolution of biological life in his 1963 paper [16, pp. 123–124].

[22] The paper would have made an excellent addition to the inaugural Interdisciplinary Workshop on the Synthesis and Simulation of Living Systems (the original precursor of the current Artificial Life conference series) held in Los Alamos, NM in the same year, 1987. Sadly, Barricelli did not attend the workshop.

5.3 Realisations of Physical Self-Reproduction

The first publications describing implementations of *physical* self-reproducing systems appeared just a few years after Barricelli's initial work on software-based (logical) self-reproduction. The two most significant authors of work on early hardware-based self-reproduction are Lionel Penrose and Homer Jacobson. Both of their projects followed Barricelli's approach; they employed systems in which the elementary units are contrived such that particular compound chains of units can catalyse the formation of further copies of the same compound. These chains, which can be very short, can therefore reproduce without the complexity of von Neumann's design. While Barricelli had successfully demonstrated the first implementation of evo-replicators in software, Penrose and Jacobson's physical implementations focused on simple standard-replicators; although, as we describe below, they both discussed how their models could be further extended to improve their evolutionary potential.

5.3.1 Lionel Penrose (1957)

Lionel Penrose (1898–1972) was a distinguished British scientist best known for his work in human genetics and intellectual disability [136]. His interests within these fields were broad, and he published on many topics beyond his core research. Over the period 1957–1959, Penrose produced three papers describing a series of designs of self-reproductive chains of mechanical wooden units [237, 233, 234].[23] His aim was to investigate very simple forms of self-reproduction that could potentially have been employed by the earliest forms of life on Earth. As his son Roger recalls, "[o]ften he illustrated his scientific ideas by means of working models. ... Frequently these models would be aids to clarifying his own thoughts as much as demonstrations for others. But there is no doubt that a major influence in driving him on in this direction was his sense of fun and sheer enjoyment in constructing things from wood and other materials" [136, p. 545].

For the purposes of these models, Penrose defined a self-reproducing structure as follows:

> A structure may be said to be self-reproducing if it causes the formation of two or more new structures similar to itself in every detail of shape and also the same size, after having been placed in a suitable environment.
>
> Lionel Penrose, 1958 [233, p. 59].

---

[23] The first of these was a brief letter co-authored by Penrose's son, the now-eminent physicist Roger Penrose [237]. The later publications were more extensive, and authored by Lionel Penrose alone [233, 234]. The Wellcome Library in London has a treasure trove of original notes, correspondence, photographs, etc. relating to Penrose's research, freely available in digitised form on their website at https://wellcomelibrary.org/item/b20218904. Items on the website are categorised hierarchically, with all of the work on self-reproduction filed under reference PENROSE/2/12/x. In the following footnotes, any references of this form can be found by navigating from the URL given above.

He did not restrict his investigation to cases of non-trivial self-reproduction in von Neumann's sense (i.e. to systems capable of an evolutionary increase in complexity via heritable mutations). However, he did set out certain standards to guide his work: that the elementary units of the system "must be as simple as possible"; that there "must be as few different kinds [of unit] as possible"; and that the units "must be capable of forming at least two (preferably an unlimited number) of distinct self-reproducing structures" [234, p. 106]. One further restriction was that a self-reproducing structure could only be built by copying a previously existing seed structure, i.e. there could be no spontaneous generation of self-reproducing structures. The logic for this final restriction was the fact that spontaneous generation was not observed in biological life [234, p. 106].[24]

The first paper described a simple system comprising multiple copies of two basic types of wooden block. The two types (which we will refer to as A and B) look deceptively simple, but were cleverly designed so that, when in a particular orientation, an A unit and a B unit could hook together in two distinct ways (A-B or B-A), as shown in Fig. 5.5.

The units were restricted to move along a horizontal channel of limited height such that they could not pass one another. In this system, if a collection of individual A and B units were randomly placed in the channel and the channel was shaken horizontally, the units would collide but would never link together. However, if a single A-B linked unit was placed in the channel before the channel was shaken, it acted as a "seed". The seed caused individual units colliding with it to assume the correct orientation to form further linked A-B units. The same result occurred with an initial B-A seed too, but the reproduction always bred true to the initial seed.

This first design served as a demonstration of how physical self-reproduction could be implemented with an extremely simple mechanism. From the experience gained with this design, Penrose identified five principles of self-reproduction which guided his later work [233, pp. 61–62]. In contrast to von Neumann's work on the logical design of self-reproducing machines, Penrose's principles very much focused on physical and energetic concerns: (1) The units must each have at least two possible states, one of which is a neutral state (in which potential energy is lowest) and other states which are associated with various degrees of activation; (2) In order to form a self-reproducing structure, kinetic energy must be captured and stored as potential energy in its constituent units (hence the constituent units are activated); (3) The activated structure or machine must have definite boundaries (unlike a crystal, for example); (4) Each activated unit must be capable of communicating its state to another unit with which it is in close contact; and (5) The chances of units becoming attached correctly can be increased by guides (or tracks, channels, etc.), which act like catalysts. These guides could be in the environment (as in the simplest system described above) or could be part of the units themselves (e.g. interlocking edges).

From this starting point, the later papers [233, 234] described a series of progressively more complicated models. The most complicated schemes that Penrose

---

[24] Of course, anyone considering the use of models of this type to explore the *origin* of life would have to relax this restriction.

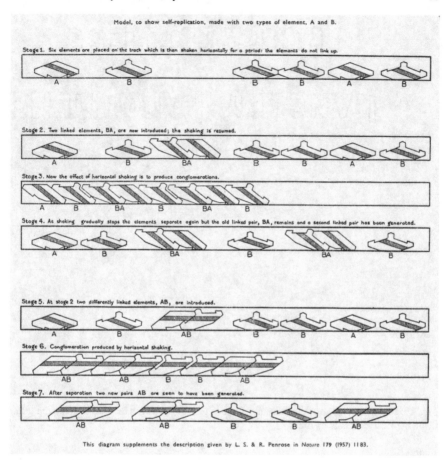

Fig. 5.5 Penrose's simple two-unit model showing replication of an introduced B-A seed unit (middle) and of an introduced A-B unit (bottom). A description of the system is provided in [237].

devised and built were inspired by the way DNA strands are copied through a process of template reproduction by base pairing (see [233, pp. 68–71] and [234, pp. 109–114]). The elementary units of these schemes were hybrids of an elaboration of the simple self-catalysing pairs of A and B units from Penrose's simplest model (which acted as the "base pairs"), together with a system of guides, passive hooks and activation and release mechanisms. These additional features allowed the formation of linear chains of units of arbitrary length and promoted the coordinated sequence of associations and disassociations required for the chains to reproduce. A schematic showing the operation of one of these more complex designs, with handwritten annotations by Penrose, is shown in Fig. 5.6. Photographs of his physical implementation of another of his more complex designs are shown in Fig. 5.7.

Penrose discussed how self-reproducing structures in a system like this might perform other actions in their environment beyond self-reproduction, dependent

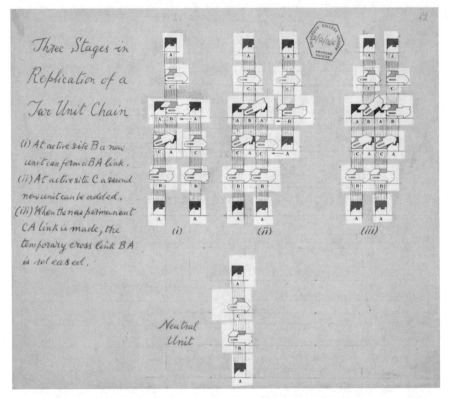

Fig. 5.6 Handwritten annotated schematic of one of Penrose's designs for a linear-chain replicator (from the Wellcome Library archive).

upon their configuration. Such functions could be subject to natural selection, he argued; moreover, mutations and recombinations could occur, allowing the structures to participate in an evolutionary process—becoming evo-replicators rather than just standard-replicators [234, pp. 112–114].

Echoing the statements made by Barricelli about the ontological status of his symbioorganisms, Penrose stated that his designs were not theoretical models but self-reproducing machines in their own right [234, pp. 114].[25] He argued that they demonstrated that self-reproduction and evolution could happen in relatively simple systems, and that work such as this could help us in understanding the early evolution of life on Earth before the emergence of complex DNA-based reproduction. The relationship between his self-reproducing machines and biological self-reproduction was further explored in subsequent publications (e.g. [235, 236]).

---

[25] Penrose did not refer to Barricelli's work in his papers, although he did become aware of it at some stage—the Penrose collection at the Wellcome Library includes a copy of a technical report that Barricelli had published in 1959 [14] (PENROSE/2/12/17/11; see footnote 23, p. 53).

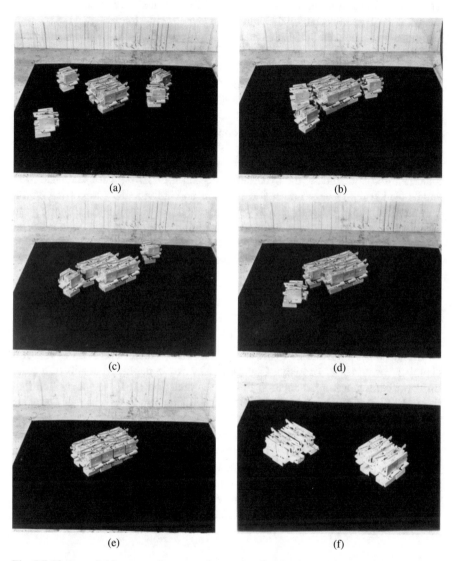

(a)

(b)

(c)

(d)

(e)

(f)

Fig. 5.7 Photographs demonstrating successive stages of replication in Penrose's physical implementation of one of his more complex self-reproduction schemes. The original four-unit seed replicator is shown in the center of (a), surrounded by four individual food units. Over (b)–(e) the food units attach to the seed in the correct places to form two copies of the original seed (e), which separate once formed (f).

Two films were produced showing Penrose demonstrating his various models.[26] The first, entitled "Automatic Mechanical Self-Replication", was made in 1958 and shown at the Tenth International Congress on Genetics in Montreal, and the second, "Automatic Mechanical Self-Replication (Part 2)", which shows the most advanced models, was made in 1961 and shown at the Institute of Contemporary Arts in London.

Penrose's work attracted widespread attention at the time. In addition to presenting numerous academic lectures and demonstrations on the topic,[27] one of the articles referred to above was published in the popular science magazine *Scientific American* [234]. Furthermore, Penrose appeared in an episode of the BBC TV series *Science International* in December 1959, which had a special feature on the topic "What is Life?", [28] and his original simple self-reproducing units were also manufactured and sold by the cybernetics organisation *Artorga* for demonstration and education purposes to universities, schools and individuals.[29] Reporting on the presentation of the first film at the Tenth International Congress on Genetics, the Montreal Gazette voiced fears that such work might potentially lead to an artificial self-replicator that got "out of control" and declared that "[t]he implications of this work are tremendous—and a little terrifying" [52].

5.3.2 Homer Jacobson (1958)

At around the same time as Penrose's publications, Homer Jacobson (1922–), a chemistry professor at Brooklyn College in New York, published details of a quite different hardware implementation of self-reproduction [151]. He suggested that the real value of this kind of model was to "call attention to the abstract *functions* inherent in the processes they represent" [151, p. 264].

The paper began by setting out what Jacobson described as the "obvious essentials in any reproducing system." These were (1) "An environment, in which random elements, or parts, freely circulate"; (2) "An adequate supply of parts"; (3) "A usable source of energy for assembly of these parts"; and (4) "An accidentally or purposively assembled proto-individual, composed of the available parts, and synthesizing them into a functional copy of its own assembly, using the available energy to do so" [151, p. 255].

---

[26] PENROSE/2/12/14 (see footnote 23, p. 53). At the time of writing, several versions of the films can be found online. Links to these can be found at http://www.tim-taylor.com/selfrepbook/.

[27] E.g. PENROSE/2/12/2, PENROSE/2/12/3, PENROSE/2/12/4, PENROSE/2/12/12 (see footnote 23, p. 53).

[28] PENROSE/2/12/9 (see footnote 23, p. 53). Other highlights of the episode included the pioneering molecular biologist Sydney Brenner discussing the genetic code, and Harold Urey describing his work with Stanley Miller on early terrestrial chemistry and the origins of life.

[29] PENROSE/2/12/11 (see footnote 23, p. 53). From the full PDF file available to download from this page, see especially pp. 100–108, 117–140, 174, 177.

While this list of requirements is fairly close to Penrose's assumptions, Jacobson was clearly thinking of a somewhat more complex, electromechanical setting for his work. He suggested that the minimum set of part types necessary to build such a system would include (1) an energy transducer, (2) an information storage medium containing some kind of plan for assembly of an organism, and (3) some kind of sensory system [151, p. 255].

The implementation was based upon a model railway track, around which two types of specially adapted locomotive cars circulated.

In the simplest version described in the paper, the track was oval and featured a number of sidings. Type A cars were equipped with electromechanical machinery inspired by the punch card, and stored the instructions necessary to direct the reproduction of a seed "organism". Type B cars were equipped with devices both for controlling points on the track (which determined whether passing cars would continue on the main track or be diverted to a siding) and for sensing if a car was nearby, and if so, of what type.

The self-reproduction process was initiated by placing a seed organism on one of the sidings. This comprised two connected cars, a type A car at the front with a type B car behind. Communicating with each other to coordinate their actions, the A and B cars of the seed organism picked out single A and B cars as they passed by on the main track, and controlled the points to create a replica two-car organism on an adjacent siding.

Jacobson successfully designed and implemented a working version of this system, with full details given in the paper. He also described some possible extensions to the system, including versions where the seed could produce more than one offspring, where the organisms could lay their own points and sidings, where the information on the organism's punch card was copied during the reproduction process rather than pre-existing in the individual A cars,[30] and where the constituent cars of "dead" (dormant) organisms could be recycled for use in further rounds of reproduction. Jacobson provided notes (in varying degrees of detail) on how these features might be implemented in theory, although the complexities involved meant that they remained unimplemented in practice.[31]

The paper includes a discussion of the complexity of the system's parts, in which Jacobson notes that "living beings are complex assemblies of simple parts, while the models are simple assemblies of complex parts" [151, p. 264]. He admitted that reproduction in his models relied upon the detailed properties of an "extremely arbitrary" environment (he viewed the environments in von Neumann's proposed

---

[30] Jacobson also mentioned the possibility of mutations arising in the genetic instructions, although he did not explore the evolutionary implications of this at any length [151, p. 274].

[31] Jacobson also remarked that a different model of self-reproduction could be built into an entirely electronic system (i.e. logical rather than physical self-reproduction). Interestingly, he cited Barricelli in the paper, although the cited papers related to Barricelli's speculations on the role of gene symbiosis in the origins of modern life. There is no indication that Jacobson was aware of Barricelli's work on software-based self-reproduction and evolution. He did not refer to Penrose's work in his paper either, although the two certainly corresponded later on—the Penrose collection at the Wellcome Library includes an offprint of Jacobson's paper with the handwritten note "Best regards, Homer Jacobson" on the front cover (PENROSE/2/12/17/8).

models of self-reproduction as equally arbitrary), and that their "unnatural (i.e. arti-
factual) quality" makes them uncomfortable for biologists to accept. In other words,
the design of the basic parts and dynamics of these environments was not guided by
any fundamental physical principles but rather by the specific intention of allowing
particular structures to reproduce. He suggested that models such as these could perhaps be "rated as to an elegance factor which increases as the environment is made
simpler, with less built-in instrumentation, and simpler parts" [151, p. 263].

The discussion ended with some consideration of the application of information
theory to compare the informational capacity of the model system to estimates of the
informational capacity of minimal forms of biological life.[32] These considerations
led Jacobson to conclude that "[his] models, information theory, and thermodynam-
ics all seem to agree that the probable complexity of the first living being was rather
small" [151, p. 268].

The year after Jacobson's paper appeared, Harold Morowitz published a brief
letter in the same journal in which he described a simple electromagnetic self-
reproducing system comprising two distinct types of part floating in water [211].
Morowitz cites Jacobson as his inspiration, although some aspects of his design are
closer to Penrose's approach.

5.4 Scientific Speculation in the 1950s

In addition to advances in theory and working models, the 1950s also saw continued speculation on the longer-term applications and impact of artificial self-
reproducing systems. Some relevant work from science fiction was already discussed in Sect. 4.1.3. Beyond this, we now look at speculative work by scientists
during this period.

5.4.1 Edward F. Moore (1956)

Perhaps the most notable speculation on applications of self-reproducing systems
from the scientific literature of the 1950s can be found in a *Scientific American* article by the American mathematician Edward F. Moore (1925–2003). Taking von

---

[32] In 1955, Jacobson had published a paper on the application of information theory to aspects of
reproduction and the origins of life [150]. In it he referred to mechanical models of reproduction,
explicitly mentioning those "conceived, in rather abstract terms, by von Neumann" [150, p. 122],
and also a "specific and simple model" designed by himself—details of which he said would be
reported elsewhere. This is presumably the model described above and reported in [151]. Interestingly, Jacobson and his 1955 paper were recently in the news when, in 2007 (over half a century
after its original publication), he decided to retract two brief passages which he had come to view
as unfounded or mistaken, and which had become much cited by creationists as evidence for the
impossibility of life arising by accident (see [152] for the retraction and [79] for coverage of the
story in the New York Times).

Neumann's work as a starting point, it proposes a research programme to design and build much more ambitious self-reproducing machines. In addition to self-reproduction, Moore's machines would be able to produce materials of economic value, which could then be harvested with exponentially increasing yields [208]. His interests were therefore squarely on the potential of self-reproducing machines as maker-replicators. Moore introduced his proposal by explaining:

> I would like to propose [a] self-reproducing machine, more complicated and more expensive than Von Neumann's, which could be of considerable economic value. It would make copies of itself not from artificial parts in a stock room but from materials in nature. I call it an artificial living plant. Like a botanical plant, the machine would have the ability to extract its own raw materials from the air, water and soil. It would obtain energy from sunlight ... It would use this energy to refine and purify the materials and to manufacture them into parts. Then, like Von Neumann's self-reproducing machine, it would assemble these parts to make a duplicate of itself.
>
> Edward F. Moore, 1956 [208, p. 118]

The proposal outlined the general aims and challenges of the research programme, with rough estimates of overall time and costs. If sufficient effort was deployed on the project, Moore thought that 5–10 years and $50–70m would be sufficient in the best case, extending to several decades and hundreds of millions of dollars if things did not go so smoothly.

Moore envisaged the machines being constructed from electromechanical parts, rather than biochemical components, because of our better understanding of their design principles. The idea was that these artificial plants would be most useful in currently uncultivated locations, starting in areas such as seashores with relatively easy access to materials and sunlight, and potentially moving to more challenging environments such as the ocean surface, deserts and the continent of Antarctica.

Noting that von Neumann had already solved the problem of the logic of self-reproduction, Moore discussed other difficulties, arguing that the necessary chemical engineering would present the greatest challenges. Reasoning that the energy requirements of manufacture would scale with machine mass, and yet energy capture from sunlight would only scale with surface area, he envisaged "small, or at least very thin" machines [208, p. 121]. These would need to be equipped with "wheels or a propeller" to enable offspring to spread and avoid overcrowding [208, p. 124].

Moore identified the most important general criterion for success as reproduction time (such as the time required for a population of artificial plants to double in number) and suggested that to be economically sensible, this would need to be at the very least faster than "the time it takes for money to double at compound interest" [208, p. 121]. No doubt wisely, he suggested that such a machine should *not* be endowed with evolutionary abilities, "lest it take on undesirable characteristics" [208, p. 122].

The article ended with some brief discussion of potential problems from perspectives of ecology, economics and society—although the one sentence given to potential ecological problems leaves plenty of scope for further elaboration.

5.4.2 Konrad Zuse (1957)

A year after Moore's article appeared, the German computer pioneer Konrad Zuse (1910–1995) published his first paper on the subject of machine self-reproduction [316]. The article was an extract from a lecture he presented at the Technical University of Berlin on 28 May 1957 on the occasion of being awarded an honorary doctorate.[33] Whereas von Neumann had developed the subject from a theoretical perspective, and Moore had proposed advanced uses of the technology but left the implementational details to be addressed in a future research programme, Zuse wished to approach the problem of self-reproduction from the perspective of practical realisation: "To me the practical problems involved in this concept are the actual bottlenecks that must be conquered" [318, p. 163]. Later in his career, in the late 1960s and early 70s, Zuse made progress in the design of electromechanical prototypes of some components of his ideas (see Sect. 6.3), but in his 1957 paper he sketched out his grand plans for how the technology might be used.

The foundation of Zuse's vision was the concept of a self-reproducing workshop, made from a sufficient diversity of manufacturing machines so as to achieve closure in the ability for the workshop to manufacture new copies of every machine. We will further discuss the topic of closure in Sect. 7.3.1. We see echoes in Zuse's approach to Butler's earlier ideas in *Erewhon* of the collective reproduction of heterogeneous groups of machines (Sect. 3.1). Zuse saw that such a factory would be able to produce other machines in addition to those from which it was itself comprised; like von Neumann and Moore, he was interested in the possibilities of maker-replicators. This being the case, he said "the question that is of greatest interest is the following: What is the simplest form of initial workshop necessary to crystallize out of it a complete industrial plant?" [316, p. 163].[34] Zuse named this idea, of a minimal self-reproducing seed out of which whole industrial plants for specific purposes could grow, the *technical germ-cell*.

Pushing the idea further, Zuse imagined that a technical germ-cell like this could be directed to make a copy of itself at a slightly smaller scale, leading to a lineage of germ-cells of progressively smaller size. The result, he envisaged, would be a microscopic germ-cell that was "not only ... the constructionally and logically simplest form, but also the smallest in space" [316, p. 164].[35] From this microscopic germ-cell, which Zuse referred to as the "real germ-cell" [316, p. 164],[36] an entire human-scale industrial plant could be manufactured by reversing the sequence of miniaturisation steps through which the real germ-cell had been created. Using this technology, he speculated, future generations of engineers might not *build* indus-

---

[33] The date of Zuse's presentation was confirmed by staff at the University Archives of the Technical University of Berlin [personal communication to TT, 14 August 2018].

[34] Quotation translated by TT from the original German text: "Die Frage, welche dann von größtem Interesse ist, ist folgende: Welche einfachste Form einer Anfangswerkstatt ist erforderlich, um aus ihr ein vollständiges Industriewerk auskristallisieren zu lassen?" [316, p. 163].

[35] "nicht nur, die konstruktiv und logisch einfachste Form ... sondern auch die räumlich kleinste" [316, p. 164].

[36] "die echte Keimzelle" [316, p. 164].

trial plants and factories but *plant* them and simply supply them with sufficient raw materials and energy with which to grow. Zuse's ideas call to mind Philip K. Dick's image at the end of his short story *Autofac* of miniature seeds of manufacturing plants being launched into space (Sect. 4.1.3). Dick's story appeared in 1955, two years before Zuse's talk. However, comments in Zuse's handwritten notebooks from 1941 confirm that he had originated the idea independently and had been thinking along these lines for a long time [315].

Zuse ends the 1957 paper with the thought that the technical germ-cell could also be used to manufacture computing devices, and that this might be a route to evolve artificial intelligence that could eventually be "able to perform ... inventions and mathematical developments ... better than man" [316, p. 165].[37]

Read in isolation, the ideas relating to the technical germ-cell set out in Zuse's paper appear somewhat far-fetched from an engineering perspective—although perhaps no more so than Moore's paper from the previous year. Nothing is said about the information and control mechanisms that would be required to drive the processes he discussed. Furthermore, regarding the process of miniaturisation Zuse envisaged to arrive at the "real germ-cell", he acknowledged that different physical principles and manufacturing methods would be applicable at different physical scales—but simply noted that this would be a significant problem to be tackled in future research [316, p. 163]. However, this was a transcript of a relatively short speech given at an honorary degree ceremony; it was not an occasion to present a long discussion of detailed technicalities. Indeed, at the end of the speech Zuse asks of his audience "Forgive me, please, if I let the imagination play farther than is usual at scientific conferences" [316, p. 165].[38] In later publications it is clear that Zuse had realistic expectations about the timescales and challenges that would be involved, and he was also serious about the enormous significance of its potential long-term applications. In the following decade he wrote at greater length and in more detail about his ideas, and also commenced work on some prototype hardware. We describe these later works in Sect. 6.3.

5.4.3 George R. Price (1957)

Another speculative essay from around the same time is of interest mostly for historical and biographical reasons, rather than for making an original contribution. Presented in the form of a fictional vision, *The Maker of Computing Machines* was written by George R. Price (1922–1975) and appeared in the pioneering computing magazine *Computers and Automation* in 1957 [240]. Price is best known for his contributions to evolutionary theory, yet he worked in various fields in his early career, including a number of years as an IBM employee in the 1960s [135].

---

[37] "Erfindungen und mathematischen Entwicklungen besser durchzuführen als der Mensch" [316, p. 165].

[38] "Verzeihen Sie mir bitte, wenn ich heute einmal die Phantasie etwas weiter habe spielen lassen, als es sonst auf wissenschaftlichen Tagungen üblich ist" [316, p. 165].

The story describes an inventor who creates a series of progressively more complex computing machines—and, later, robots—incorporating mechanisms such as self-repair, goal-based behaviour, associative learning and long-term planning. The advanced robots eventually learn how to make some of the earlier computer designs themselves, but they also acquire some unsavoury behaviours such as torturing and killing their own kind. The inventor had wanted his robots to develop qualities of altruism, benevolence and cooperation, but he found it too difficult to codify these into explicit goals. So he equipped the robots with simple built-in goals such as keeping themselves in repair and seeking fuel, and provided them with a mechanism by which they could develop more complex derived goals by themselves. The more complex goals would enable the robots to achieve their built-in goals in creative ways, based upon their experiences as they explored and learned about their world. However, the unsavoury behaviours that were observed were quite the opposite to the inventor's intended outcome.

In this short story, Price pinpoints a key issue facing AI designers today: the problem of how to prevent learning systems from developing unwanted behaviours. In present-day debates about the dangers associated with advanced AI this has become known as the *value alignment problem* (see Sect. 7.3.4). The idea of creating intelligent machines through biologically-inspired learning mechanisms had been discussed by Alfred Marshall ninety years earlier (Sect. 3.2), and the theme of technology out of control had been raised by Samuel Butler at around the same time as Marshall (Sect. 3.1) and by many early-twentieth century authors (Sect. 4). Still, Price's story is an interesting example of a contribution from an early theoretical biologist. In highlighting the difficulties in developing cooperative and altruistic behaviour, it also anticipates Price's later work in theoretical biology on kin selection.

5.5 Self-Reproduction in the Cybernetics Literature

The late 1940s saw the birth of the scientific field of *cybernetics*, which sought a common understanding of principles of control and communication in animals and machines. Among those working in the area, there was a common view of intelligence as a *search problem*, and parallels were drawn between the processes of lifetime learning and evolution. Examples can be found, among others, in W. Ross Ashby's notion of *intelligence amplifiers* [7] and in Alan Turing's early work on artificial intelligence [295].[39] In addition to this general interest in evolution applied to machines, some of the leading cyberneticists specifically discussed the idea of self-reproducing machines. In particular, Norbert Wiener, Gordon Pask and W. Ross Ashby all published work on self-reproduction and evolution in the early 1960s.

---

[39] Turing's first published thoughts on the idea of evolution as a search process in the context of machine learning appeared in a 1948 research report entitled *Intelligent Machinery* [294, p. 18]. The director of his laboratory at the time was none other than Sir Charles Galton Darwin, grandson of Charles Darwin. He was unimpressed by Turing's report, dismissing it as a "schoolboy essay" [67].

Wiener's influential book *Cybernetics, or control and communication in the animal and the machine*, first published in 1948, was supplemented with two additional chapters in the 1961 second edition: one of the new chapters was entitled *On Learning and Self-Reproducing Machines* [311].[40] In this, he discussed the relationship between lifetime learning and evolution; while the majority of the chapter is devoted to lifetime learning systems, in the last few pages Wiener turns to the subject of self-reproduction.[41] He describes the design of an example logical self-reproducing system in the form of a particular kind of electronic circuit that could automatically imitate the behaviour of a given second circuit.

However, Wiener's proposal is not an entirely satisfactory example because the thing that is being reproduced—the behaviour of the given circuit—is not in any way playing an active role in its own reproduction. Although the reproduction process was entirely automated, it relied upon the existence of additional electronic circuitry in order to bring about the reproduction of the reference behaviour. As we have seen in other work described in this chapter, the issue of how much of the process of reproduction should be directed by the self-replicator itself, and how much should rely upon very specific features of its operating environment, was a topic discussed by others researchers of the time, including von Neumann (Sect. 5.1.1), Jacobson (Sect. 5.3.2) and Ashby. We return to discussion of this issue in Sect. 7.3.

Also published in 1961 was Pask's *An Approach to Cybernetics* [229], which included a chapter entitled *The Evolution and Reproduction of Machines*. Pask considered the problems involved in creating artificial evolutionary systems that exhibit on-going, open-ended evolutionary activity; his focus was therefore very much upon the possibilities of evo-replicators. He noted that if, over time, the environment experienced by any one machine becomes increasingly determined by other evolving machines in the system, this will produce an "autocatalytic" effect that avoids the need for the system designer to build a drive for evolutionary trends into the system's reward structure [229, pp. 101–102]. He also argued that the successful evolution of a species of machine in a competitive environment was likely to involve the sequential development of a hierarchy of levels of description (or "metalanguages") used in its genetic or control structures, with which the machines would encode information about their own design and their relationship with the environment [229, p. 101].[42]

A somewhat different approach to the subject was taken by W. Ross Ashby in his 1962 paper *The Self-Reproducing System* [8]. Like Penrose and Jacobson before him, Ashby emphasised that self-reproduction is a function of the *interaction* between the form that is reproduced and the environment within which the form exists as a subsystem. Considering self-reproduction processes in general, he argued (as did von Neumann and Jacobson) that the allowed complexity of the building blocks is essentially an arbitrary decision. To demonstrate this, he gave a string of exam-

---

[40] Burks states that Wiener and von Neumann interacted when developing their respective conceptions of cybernetics and the theory of automata [10, p. 364].

[41] In these last few pages Wiener uses the term *self-propagation* rather than *self-reproduction*, in contrast to his use of the latter term in the opening pages of the chapter and indeed in the chapter title.

[42] We will return briefly to this point in Sect. 7.3.

ples of real-world systems that fit his basic definition of self-reproduction. In most of these, the form being reproduced is relatively simple, and the process of reproduction is largely due to the particular (complex) environment in which the form exists. One of his more whimsical examples is a yawn, which reproduces in a suitable environment of people. Given this, he argued that it is the particular properties of the terrestrial environment on Earth that distinguish the processes of biological self-reproduction and evolution from these other, less interesting, examples. As we already saw with Wiener's work, this question of how much of the complexity of the process of reproduction resides within the environment rather than within the self-replicator itself is indeed an important issue, and we return to it in Sect. 7.3. It is also the case that many of the examples of self-reproducing systems offered by Ashby were standard-replicators, which were not capable of heritable mutations as required by von Neumann's test for "interesting" self-reproduction (Sect. 5.1.1); that is, they were not evo-replicators.

Developments in cybernetics around this period were not confined to the United States and Europe. In the Soviet Union during the early 1950s the field had initially been greeted with scepticism, being regarded by some as a "reactionary pseudoscience" that served the interests of the bourgeoisie by reflecting its desire to "replace potentially revolutionary human beings with machines" [144, p. 299]. By the late 1950s, however, attitudes had changed, as attested to by the establishment in 1959 of a Science Council for Cybernetics by the USSR Academy of Sciences [144, p. 299].

In 1961, the Russian polymath Andrei Nikolaevich Kolmogorov (1903–1987) presented a well-attended seminar entitled "*Automata and Life*" to the Faculty of Mechanics and Mathematics at Moscow State University. In this and two subsequent talks in 1962 he explored the implications of materialism, and of the nascent field of computer science, for our understanding of life and mind. The talks in 1962 were entitled "*Life and thinking as special forms of the existence of matter*" and "*Cybernetics in the study of life and thinking.*" Some general information regarding the circumstances of these talks is provided in [298, p. 497].

Kolmogorov prepared an extended abstract for *Automata and Life*, which was later published in various different editions.[43] The content of the three lectures also formed the basis of a second paper, *Life and thinking as special forms of the existence of matter*, which was also published in several different forms.[44]

---

[43] It first appeared in 1961 in [165] (in Russian). An English translation of this article was produced by the U.S. Department of Commerce's Joint Publications Research Service the following year [166]. Some of the later Russian versions were prepared and edited by Natalya Grigorevna Khimchenko (née Rychkova)—who worked in Kolmogorov's research group in the early 1960s—to make them suitable for a wider audience. Useful information on the various versions of the talk material, and the differences between them, is provided by Khimchenko at http://vivovoco.astronet.ru/VV/PAPERS/BIO/KOLMOGOR/KOL_REP.HTM (in Russian).

[44] Quotations from *Life and thinking as special forms of the existence of matter* provided in this chapter are based upon the original version published in 1964 [167] (in Russian), with English translation by TT aided by Google Translate. An English translation based upon a later version [169] was published by NASA [170], although this is somewhat abridged compared to the original version.

In these talks Kolmogorov suggested that the age of space travel brought with it the prospect of encounters with extraterrestrial intelligent beings. This presented a pressing need for understanding the concept of life in more general terms, abstracted from the specific chemical details of life on Earth. Likewise, the advent of the computer age created an urgent need to conceptualise thought and cognition in more general terms.

Kolmogorov opened *Automata and Life* with three questions:

Can machines reproduce their kind, and in the course of such self-reproduction can progressive evolution take place leading to the creation of machines of a higher degree of perfection than the originals?

Can machines experience emotions?

Can machines have desires and can they pose for themselves new problems not put to them by their constructors?

A. N. Kolmogorov, *Automata and Life*, 1961 [165, p. 3][45]

We can see in these questions that his focus was specifically on evo-replicators. In these talks aimed at a wide audience, Kolmogorov did not delve too deeply into technical responses to these topics. However, he did point out that, within the framework of a materialist worldview, it must be admitted that there are no fundamental arguments against a positive answer to these questions [167, p. 53]. In other words, if biological organisms can do these things, then it should, in principle, be possible for machines to do them too.

Kolmogorov cautioned that the current cybernetics literature displayed both many exaggerations and many simplifications [168, p. 25]. Furthermore, he observed that research on understanding human behaviour was focused on the most simple conditioned reflexes, on the one hand, and on formal logic, on the other hand [168, pp. 27–28]. The vast space between these two extremes in understanding the architecture of human behaviour, Kolmogorov noted, was not being studied in cybernetics at all. Hence, he suggested, the kind of developments he was discussing may take longer to come to fruition than many people might expect.

Taking early attempts at musical composition by computers as a specific example, Kolmogorov warned that in order to properly simulate or build living beings, we need to understand the source of their internal desires and not just purely external factors; to design a computer that can generate interesting music, we need to understand the difference between living beings in *need* of music and beings who do not need it [168, p. 26]. Having argued that we could, in theory, fully understand

---

[45] English translation from [166, p. 1]. According to Natalya Khimchenko, the original talk abstract prepared by Kolmogorov phrased the second question as "*Can machines think and experience emotions?*" (see website listed in footnote 43, p. 66). In a later version of the text that appeared in the 1968 publication *Kibernetika Ozhidaemaya i Kibernetika Neozhidannaya* (Cybernetics Expected and Cybernetics Unexpected), the questions are stated as follows: "*Can machines reproduce their own kind, and can there be a progressive evolution in the process of such self-reproduction, leading to the creation of machines that are significantly more advanced than the originals? Can machines experience emotions: rejoice, be sad, be dissatisfied with something, want something? Finally, can machines set themselves tasks not assigned to them by their designers?*" [168, p. 14] (English translation by TT aided by Google Translate).

our own design principles, Kolmogorov suggested that we should not be afraid of creating automata that imitate our own abilities [168, p. 31] or, indeed, of creating automata just as highly organized but very different from us [168, p. 15]. Rather, we should take great satisfaction in the fact that the human race has developed to the point where we can create "such complex and beautiful things" [168, p. 31].

It is interesting to note the parallels between Kolmogorov's thoughts in this article and those of Ada Lovelace in her comments on Charles Babbage's *Analytical Engine*, written over a century earlier in 1843.[46] As Kolmogorov cautioned against hype in the promise of cybernetics, so too had Lovelace cautioned against "exaggerated ideas that might arise as to the powers of the Analytical Engine" [204, p. 722]. However, Lovelace had gone on to state that the "Analytical Engine has no pretensions whatever to *originate* anything. It can do whatever we *know how to order it to perform*" [204, p. 722] (original emphasis). In contrast, in answering his third question in the affirmative, Kolmogorov had suggested that it should, in principle, be possible to build machines that *can* originate new goals. Finally, Lovelace famously suggested that "the engine might compose elaborate and scientific pieces of music of any degree of complexity or extent" (but only if "the fundamental relations of pitched sounds in the science of harmony and of musical composition" could be codified into a suitable sequence of operations) [204, p. 694]. As previously stated, Kolmogorov was less interested by the prospect of the purely algorithmic generation of music, and wanted instead to understand the design principles that instil human beings with the *desire* to compose music. We will return to the topic of how machines might develop their own purposiveness and internal desires in Sect. 7.3.4.

*   *   *

To summarise what we have discussed in this chapter, the 1950s witnessed the attainment of the third and final step in the historical development of the idea of self-replicator technology (Sect. 1.6)—the arrival of the first examples of actual implementations of self-replicators, both in software and in hardware. As we have seen, this was accompanied by the emergence of a new concept: the use of self-replicator technology in the design of universal manufacturing machines, i.e. *maker-replicators*.[47] Having covered the attainment of the three major steps in the development of thinking about self-replicator technology, in the next chapter we present a brief review of more recent developments with standard-replicators, evo-replicators and maker-replicators, from the 1960s to the present day.

---

[46] We met Babbage and Lovelace briefly at the start of Chap. 3.

[47] In addition to the work of von Neumann, Moore, Zuse and others described in this chapter, the idea of a maker-replicator was also present in some of the later sci-fi stories we mentioned in Sect. 4.1.3, including Philip K. Dick's *Autofac* (1953) and Robert Sheckley's *The Necessary Thing* (1955).

Chapter 6
More Recent Developments: Signposts to Work from the 1960s to the Present

The blossoming of theoretical and practical work from the late 1940s to the early 1960s, described in Chap. 5, continued to gather pace as the 1960s progressed. The period from the 1960s to the present day has witnessed significant developments in the field, and the work has branched into a variety of novel application areas. Most of these developments are well described in existing publications, so our detailed review of the early history of the field ends here. In this chapter we describe the general nature and focus of this more recent work, and provide references to other sources that review these developments in detail.

One of the most comprehensive reviews of work in this period, with an emphasis on hardware implementations, is provided by Robert Freitas and Ralph Merkle in their book *Kinematic Self-Replicating Machines* [119]. This covers developments with all three kinds of replicator (standard-, evo- and maker-) but particularly focuses on maker-replicators. A more concise overview, with emphasis on work in software covering all three kinds of replicator, is provided by Moshe Sipper in [263]. Both of these publications include useful diagrammatic lineages of work in this area from the 1950s onward ([263, p. 238], [119, p. xviii]). Another excellent general review of the area, covering both hardware- and software-focused work involving all three kinds of replicator, is provided by Michele Ciofalo in [62]. Finally, Matthew Moses and Gregory Chirikjian provide a very recent review, mainly focused on physical replicators, which covers developments that have occurred after the publication of Freitas and Merkle's earlier review right up to the year 2019 [212].

In the following subsections we signpost some of the broad trends that have developed in theoretical explorations and in software and physical implementations of self-reproducing systems. These developments in scientific and engineering theory and practice have been accompanied by continued interest in the idea of self-reproducing systems in science fiction. Some of the most notable examples from 1960s sci-fi include works by Poul Anderson [2], Stanislaw Lem [186], Fred Saber-

© Springer Nature Switzerland AG 2020
T. Taylor, A. Dorin, *Rise of the Self-Replicators*,
https://doi.org/10.1007/978-3-030-48234-3_6

hagen [256] and John Sladek [265]; examples from more recent decades are too numerous to list.[1]

6.1 Theoretical and Philosophical Work

Von Neumann's foundational studies, described in Sect. 5.1.1, laid the groundwork for many further theoretical developments. Much of the relevant work from the 1960s and 70s took place in the field of *automata theory*. Many of these studies continued to use cellular automata, or closely related models, as a simplified platform for implementation. An early review of these developments, written by computer scientist and neuroscientist Michael A. Arbib, appeared in the proceedings of *Towards a Theoretical Biology*—an influential conference series in the late 1960s [4]. Arbib's review highlighted topics such as what he referred to as the *fixed point problem of components*; that is, ensuring that a self-reproducing system is able to manufacture a copy of each of its constituent parts. We will return to this topic, and related issues concerning *closure* in self-reproducing systems, in Sect. 7.3.1. The first part of Arbib's discussion concentrated on the design principles of standard-replicators. This was followed by an exploration of issues relating to evo-replicators, the origin of life and real-world complexities such as dealing with noise and interaction with a rich environment. More recent reviews of work from this period can be found in [119, ch. 2] and [263].

Among Arbib's many other works of interest from around this time was a paper entitled "*The Likelihood of the Evolution of Communicating Intelligences on Other Planets*," published in 1974 [5]. Speculating on the technologies that intelligent species might utilise for interstellar communication, Arbib suggested that, while most discussion up to that point had assumed radio communication, another possibility would be the use of self-reproducing machines [5, pp. 65–66]. He envisaged that the devices could be directed to "reproduce every time they travel a constant distance ... to yield a sphere moving out from the home planet with a constant density of these ... machines" [5, p. 66]. Of course, the idea of a self-reproducing spacecraft had first been proposed 45 years earlier by Bernal (Sect. 4.2.1), but Arbib's suggestion placed more emphasis on the potential of self-reproducing technology for exponential growth in numbers. This potential—which is a property of each kind of replicator including the basic standard-replicator—was utilised in Arbib's vision to achieve (at least in theory) omnidirectional communication without loss of signal strength.[2]

---

[1] For additional references, see the partial—yet extensive—list of self-reproducing machines in fiction on Wikipedia (https://en.wikipedia.org/wiki/Self-replicating_machines_in_fiction).

[2] More recently, physicist S. Jay Olson has employed the same property in a proposed mechanism that might be used by advanced civilisations to aid their rapid expansion across intergalactic distances. The scenario involves the release of a wave of "expander" probes that "are designed to reproduce themselves and adjust their velocity slightly at pre-determined intervals, so that the expanding sphere of probes maintains a roughly constant density" [221, p. 5].

A few years later, mathematical physicist Frank J. Tipler used the idea of self-reproducing machines to argue that extraterrestrial intelligent species do not exist [288, 290, 289, 291].[3] Inspired by von Neumann's theoretical concept of a self-reproducing universal constructor, Tipler suggested that any intelligent species engaging in interstellar communication would "eventually develop a self-replicating universal constructor" [288, p. 268]. This technology would be employed, he argued, not just for interstellar communication (as suggested by Arbib) but also for interstellar travel to explore and colonise the galaxy. He referred to such spacecraft as *von Neumann probes* [288, p. 276].

Tipler's line of reasoning utilised not just the self-reproductive capabilities of von Neumann's architecture (which allowed a cost- and time-efficient means of exploring the galaxy) but also its capacity for universal construction—that is, its abilities as a maker-replicator. The key point, he explains, is that "once a von Neumann machine has been sent to another solar system, the entire resources of that solar system become available to the intelligent species that controls the ... machine; all sorts of otherwise-too-expensive projects become possible" [288, p. 270] Furthermore, "in a fundamental sense a von Neumann machine cannot become obsolete ... [because it] can be instructed by radio to make the latest devices after it arrives at the destination star" [288, p. 271]. Having set out the case for the use of self-reproducing spacecraft for interstellar travel by intelligent species, he went on to utilise the *"where are they?"* argument to conclude that such species did not exist.[4]

More recently, in discussing ways in which self-reproducing probes could be used to allow humans to colonise other planets in the age of superintelligent AI, Max Tegmark invoked a different use for a probe with universal construction capabilities. In assuming the existence of superintelligent AI with access to far more advanced technology than anything that looks even remotely possible today, the idea pushes credibility to the very limits. In Tegmark's scenario, the humans would not join the probes on their interstellar journeys. Instead, once the probes had arrived on a new planet and prepared it for our coming, they would establish a superintelligent AI (perhaps with the aid of information transmitted from the mother civilisation) which would then construct a human colony in situ by constructing embryos, or even adult humans, using nanoassembly techniques [284, p. 225].

Returning to the more general and down-to-earth landscape of work on the theory of self-reproducing systems in recent decades, a significant development was the establishment of the field of *Artificial Life* (ALife) in the late 1980s.[5] This is a

---

[3] The first three of these papers were published in the *Quarterly Journal of the Royal Astronomical Society*, and the fourth, a shorter summary of the first three, was published in *Physics Today*.

[4] As noted by Tipler, the *"where are they?"* argument had been employed by others before him (but without the focus on self-reproducing spacecraft); its origin is generally attributed to the physicist Enrico Fermi (see [257, p. 495]). However, as Tipler states in [290, pp. 136–137], the same argument is apparent in a seventeenth century work by none other than Bernard de Fontenelle, whom we met in Sect. 2.1.

[5] The field was born out of a 1987 workshop organised by Christopher G. Langton of the Los Alamos National Laboratory, NM [184]. That and subsequent workshops have now developed into an annual conference series, overseen by the *International Society for Artificial Life* (http://www.alife.org).

discipline that brings together computer scientists, biologists, ecologists, complex systems scientists, philosophers and others united in an interest in *synthesizing* and *simulating* living systems in non-biological media, including software, hardware and "wetware" (molecular systems).

To highlight just one of the interesting early works from the ALife field, J. Doyne Farmer and Alletta d'A. Belin published a paper in 1991 entitled *Artificial Life: The Coming Evolution* [109], which we quoted from at the start of Chap. 1. The paper argued that reproducing and evolving artificial lifeforms could be expected to emerge within fifty to a hundred years. In addition to providing another good review of work on self-reproducing systems in the late 1980s (with a particular focus on software systems), the paper also discussed the possibility that artificial life might evolve through non-Darwinian processes. The authors considered the potential of artificial lifeforms to accelerate the rate of evolution of their physical design by modifying their own genetic material. Farmer and Belin regard this as a kind of Lamarckian evolution [125], i.e. a process by which, in contrast to Darwinian evolution, beneficial characteristics acquired during an individual's lifetime are passed on to the individual's offspring. Indeed, the process discussed in their paper goes beyond what is normally considered as Lamarckian evolution because it involves not just the inheritance of acquired characteristics but also the *intentional* self-modification of the species by the species itself.

Farmer and Belin were by no means the first authors to explore these possibilities. The idea of self-designing machines was a common theme in the early sci-fi stories discussed in Sect. 4.1.3. And within the scientific community, Richard Laing had already demonstrated in the 1970s that Lamarckian evolution could be achieved in a simple automaton model by a process of reproduction by self-inspection [177, 178, 180]. We will return to this topic in Sect. 7.1.4.

Looking at current ALife research, there is an emerging focus in the field on the topic of *open-ended evolution*—the capacity apparent in the biological world to continually evolve, to discover new tricks and to increase its maximum complexity over time in a seemingly never-ending way [281, 226]. No artificial evolutionary system to date exhibits anything like this capacity; instead, after an initial burst of activity they tend to reach a more or less stable state beyond which no further innovations are observed. In contrast to work on maker-replicators, those studying ALife evo-replicator systems are keen to understand and unleash the creative power observed in biological evolution.

As mentioned in Sect. 1.4, open-ended evolution has recently been described as a "grand challenge" for the field [269]. Some view it as a promising route for producing agents with highly sophisticated artificial intelligence,[6] even including superhuman-level artificial general intelligence (AGI)[7]; this is, of course, merely the latest manifestation of the core idea behind much of the work we described in Chaps. 3–5 which dates back as far as the 1860s. Related to this, open-ended evolu-

---

[6] A curated series of video interviews with leading current AI researchers on the potential of evolutionary techniques is available at https://www.sentient.ai/labs/experts/.

[7] For an interesting discussion of the possibility of evolving AGI, see [262]. For a longer and more general discussion of the prospects for AGI and its implications for humankind, see [284].

tion could be a route whereby evo-replicators develop the ability to act according to their own ends and desires, beyond any original goals set for them by their human designers. We return to this topic in Sect. 7.3.4.

6.2 Software Implementations

From the 1960s onwards, when computers became more widely available as a tool for scientists and engineers, many more researchers started implementing self-replicators in software.

Following von Neumann's original cellular model and the developments in automata theory referred to in the previous section, there has been much further work on cellular automata models and implementations of self-reproduction. Most of this work, particularly in the earlier years, investigated design issues in the process of self-reproduction itself, and ways to make the systems perform other tasks in addition to self-reproduction—that is, the focus of this work has generally been on software standard- and maker-replicators rather than evo-replicators. Good overviews of this area can be found in [247] and [248].

In contrast, evo-replicators have been the main focus of another branch of software-based work which investigates the evolution of self-reproducing computer programs. Much of this work has occurred within the field of Artificial Life, where the approach was made popular in the early 1990s by the *Tierra* system developed by ecologist Tom Ray [244]. In Tierra, populations of computer programs compete for space and CPU time to build copies of themselves within the computer's memory. The copying process is subject to some noise so that the copies are not always perfect and small variations start to appear in the offspring programs. Because memory space and CPU time are limited, programs best adapted for survival and reproduction in this environment persist by natural selection, and less well-adapted programs die out. Ray observed not only the evolution of increasingly faster, more efficient self-reproducing programs but also the emergence of various ecological interactions. For example, small parasitic programs were seen to evolve which were unable to reproduce unaided but, instead, hijacked the code of neighbouring programs to copy themselves. This line of research is still thriving today, especially in work using the *Avida* software platform [220] as a test bed for studies in experimental digital evolution (see Fig. 6.1). These systems are described at length in many sources (e.g. [11, pp. 195–223], [155, pp. 215–274], [283, pp. 51–57]). As mentioned in the previous section, a current focus of research in this area is in developing an understanding of how to build software evo-replicator systems with the capacity for open-ended evolution.

Partially overlapping with these approaches, the field of *Artificial Chemistries* encompasses a variety of approaches to modelling life processes such as self-reproduction and evolution; see [11] for a comprehensive recent review. A branch of this field that focuses on interactions at the ecological level is *Artificial Life Ecosystems*, described in [11, pp. 163–165] and [92].

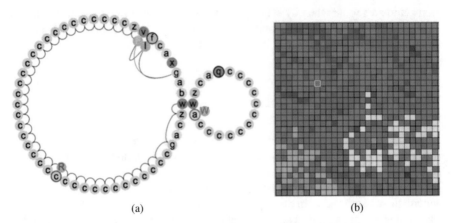

(a)   (b)

Fig. 6.1 (a) Schematic of a representative self-reproducing computer program (a digital organism) in *Avida*, and (b) example view of the two-dimensional environment in which the organisms live. Organisms have a circular genome that is read sequentially to generate behaviour, and each letter in (a) represents a single computational command from the available command set. The organism is shown in the process of creating a copy of itself. Each organism lives in a single square in the environment. The different colours in (b) represent different types of organism. In practice, environment sizes can be much larger than that shown in (b), accommodating tens of thousands of organisms.

Looking forward, it has been suggested that the process of standardisation of web technologies now presents the prospect of using the web as a globally distributed environment in which evolving software agents might find a persistent home where they could thrive "in the wild" [280].

The work outlined above has a particular focus on modelling processes of evolution, self-reproduction and related aspects of biological systems. In addition, a vast body of work has developed that uses software-based evolution primarily as an optimisation technique. The history of this work, which comes under the general name of *evolutionary computation*, has been described in various sources (e.g. [3], [207]).[8]

In a less salubrious line of development, the 1970s witnessed the emergence of computer viruses [275]. One of the first examples of a worm that spread via the Internet, causing widespread damage and attracting the attention of the mainstream media, was Robert Morris' *Internet Worm of November 1988* [82]. A good review of the history of computer viruses can be found in [268].

Much has been written about the developments described above, in the works we have mentioned and elsewhere. We will therefore leave our review of software self-replicators here and turn our attention to recent progress in the implementation of physical self-reproducing systems.

---

[8] Many of the most notable early papers in evolutionary computation, including some early work on artificial life ecosystems and papers by Nils Barricelli (Sect. 5.2.1), have recently been republished in a single volume [111].

6.3 Physical Implementations

Over the last sixty years there have been many advances in physical self-replicating systems, both at the macro-scale and at the molecular scale. A full discussion of many of the developments described in this section, and references to a wide variety of other related projects, can be found in [119], which covers work up to 2004. A good review of work over the period 2004–2019 can be found in [212].

Penrose's early work on self-reproducing blocks (Sect. 5.3.1) has inspired a lineage of further studies, ranging from systems based upon magnetic [32, 301] or electromechanical [129] units to those employing more complex programmable robotic units [273, 319]. These works have generally focused upon systems that can produce exact copies of themselves (i.e. standard-replicators), although some could in theory transmit heritable mutations and thereby act as evo-replicators given sufficient time and raw materials. However, the time and rather specialised environments required for these systems to produce their offspring mean that a great deal of further research and development is required to produce a physical self-reproducing machine that exhibits any significant evolutionary behaviour in practice.

At the same time, other researchers are exploring how additive manufacturing technology (3D printing) might be employed for the fabrication of complete robotic systems. While the technology is not yet at the stage of allowing the unassisted printing of a full robot in the general case (including all the different materials required for its electronics, actuators, power source, etc.), work is rapidly progressing in that direction (e.g. [20, 162, 194, 185]). In the meantime, a growing number of projects are investigating the use of "human-in-the-loop" 3D printing systems to partially automate the process of evolving new robot designs (e.g. [190, 140, 249, 36, 35, 132]). These lines of development might ultimately lead to the creation of fully autonomous self-reproducing and evolving systems (e.g. [30, 147]).

These developments are closely associated with the more general field of *evolutionary robotics*, which emerged in the early 1990s alongside Artificial Life. While many interesting advances have come out of this field, the majority of work tends to focus not on self-reproduction but on the evolution of controllers which are then implanted into robots of fixed physical form.[9] A good review of the field can be found in [300].

In the 1950s and early 1960s, Homer Jacobson (Sect. 5.3.2) and Norbert Wiener (Sect. 5.5) had both suggested that a self-replicating system could be developed using electronic circuits. Forty years later, in the late 1990s, this idea was realised in the *Embryonics* project, which aimed to develop an architecture for highly robust integrated circuits with the capacity for self-repair and self-replication [196].

As mentioned in Sect. 5.4.2, Konrad Zuse had started thinking about the potential of self-reproducing machines in the 1950s. His main interests lay in the possible

---

[9] The related fields of *swarm robotics* and *self-reconfigurable modular robotics* involve systems whose physical form can change, although self-reproduction is not a common concern in these fields either.

uses of maker-replicators, although he also discussed the evolutionary potential of evo-maker-replicators. After a decade of working on other projects, he returned to the topic in the second half of the 1960s.

In 1967, Zuse published an article setting out some more detailed ideas for the implementation of the technical germ-cell that he had first discussed a decade earlier [317]. We devote some time to it here because it is an extension of the work we described in Sect. 5.4.2, and because it has not been widely discussed elsewhere.

In the paper he discussed the biological cell as the inspiration for his idea of a technical germ-cell, providing an incentive for "a project that at first seems absurd to continue given the state of the art" [317, p. 58].[10] Zuse introduced the concept of the *Rahmen* (frame) of a self-reproducing system, being "the environment in which the systems are viable" [317, p. 59],[11] including all the external facilities required to provide the system's inputs and to accommodate its outputs. The inputs to the Rahmen might include raw materials, energy, information, prefabricated parts, tools, etc. [101, pp. 91–105]. He saw the degree of autonomy of a self-replicator as depending upon the complexity of the Rahmen required for its operation [317, p. 60]. The concept of a Rahmen is therefore a formalisation of the question of how much a "self"-replicator relies upon properties of its environment to achieve reproduction. As we have seen previously, this issue was discussed by von Neumann, Penrose and Jacobson before him, and we will return to the issue in Sect. 7.3.

Zuse suggested that progress could be made in the challenge of creating more autonomous self-replicators by making radical simplifications in the standardisation of individual parts, thereby reducing the inventory of parts required from the Rahmen [317, p. 61]. Regarding the question of information and control of the process, he suggested that these systems could be kept external to the self-reproducing system itself and supplied as part of the Rahmen [317, p. 63]. Zuse acknowledged that this would raise the question of the extent to which the resulting system could be called *self*-replicating, but nevertheless he suspected that this would be the most practically useful design approach. This exemplifies Zuse's focus on the manufacturing and construction aspects of the problem over the information and control aspects, which was in many ways the opposite of von Neumann's approach. In the paper Zuse also discussed some of the potentially transformational uses of the technology, not only on Earth but also, in the distant future, for space travel and exploration. The paper ends with an appeal that, although these ideas seem "a bit fantastic … we must have the courage to include such possibilities in our considerations" [317, p. 64].[12]

Over the following years Zuse began building an automatic assembly machine, the SRS72, as a starting point for a self-reproducing system [100]. His plan was to simplify the practical difficulties of the system as far as possible by employing a modular design built from standardised parts. However, it appears that the

---

[10] "… ein Projekt, welches zunächst dem Stand der Technik nach absurd erscheint, weiterzuverfolgen" [317, p. 58].

[11] "die Umwelt dar, innerhalb deren die Systeme lebensfähig sind" [317, p. 59]

[12] "… noch etwas phantastisch erscheinen, jedoch müssen wir den Mut haben, auch solche Möglichkeiten in unsere Betrachtungen einzubeziehen" [317, p. 64].

machine was not completed to a working state, and Zuse abandoned the project in 1974 for unknown reasons [100]. The art conservator Nora Eibisch has recently written a book (in German) describing Zuse's work on the SRS72 [101]; a more limited source of further information in English can be found in Zuse's autobiography [318].[13]

Zuse's work brings to mind J. D. Bernal's conception of self-reproducing spacecraft for interstellar exploration (Sect. 4.2.1). From the 1970s onward there has been a wide variety of further developments in this area (e.g. [96, pp. 194–204], [222], [19, pp. 578–586]). In 1979, Freeman Dyson set out a series of thought experiments describing how various kinds of maker-replicators could be used to transform desert regions on Earth and to terraform other planets [96, pp. 197–203]. Dyson noted that the exponentially increasing scale of operation, which was a common feature of his examples and comes about without human intervention once the first self-replicator has been set in motion, elicited a sense of getting "something for nothing":

> The paradox forces us to consider the question, whether the development of self-reproducing automata can enable us to override the conventional wisdom of economists and sociologists. I do not know the answer to this question. But I think it is safe to predict that this will be one of the central concerns of human society in the twenty-first century. It is not too soon to begin thinking about it now.
>
> Freeman Dyson, *Disturbing the Universe*, 1979 [96, p. 200]

It is fair to say that there is still no conclusive answer to Dyson's question, although it remains as relevant today as it was when he raised it forty years ago. Echoing Zuse's idea of a *technical germ-cell* (Sect. 5.4.2),[14] Dyson went on to discuss extending von Neumann's work by going beyond what he called the "unicellular level" (i.e. a single monolithic machine) to build a "germ cell of a higher organism" out of which could arise "descendants of many different kinds [that] function in a coordinated fashion" [96, p. 202]. He suggested that an analysis is required of the minimum number of conceptual components required to build a system that can act as such a germ cell; this is related to Arbib's fixed point problem of components (Sect. 6.1)—we will further discuss this topic in Sect. 7.3.1.

The most substantial exploration to date of self-reproducing technology for the exploration and exploitation of other planets was an extended study by NASA in 1980 ([119, pp. 42–51]) (see Fig. 6.2). The team that conducted the study was led by Richard Laing, whose earlier theoretical work on reproduction by self-inspection we mentioned in Sect. 6.1. Another participant was Robert Freitas, co-author of the book *Kinematic Self-Replicating Machines* highlighted at the start of this chapter. In their end-of-project report, the team considered potential long-term outcomes of such research, together with philosophical, ethical and religious questions that arose from it ([118, pp. 240–249], [179]).

The fear that such technology might become out-of-control and ultimately pose a threat to the future of humanity was as real a concern for these authors as it had

---

[13] Additional sources of information include the Konrad Zuse Internet Archive (http://zuse.zib.de/) and the website of the Deutsches Museum (http://www.deutsches-museum.de/de/ausstellungen/kommunikation/informatik/filme/montagestrasse-srs72).

[14] However, Dyson does not cite Zuse in his discussion.

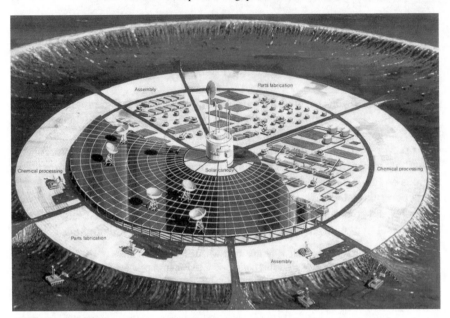

Fig. 6.2 Concept art for a self-growing lunar factory—one of the ideas explored in NASA's 1980 study of self-reproducing technology for space applications.

been for Samuel Butler over a hundred years earlier (Sect. 3.1). Although the primary focus of the study was on maker-replicators, the report suggested that "any machine sufficiently sophisticated to engage in reproduction in largely unstructured environments and having, in general, the capacity for survival probably must also be capable of a certain amount of automatic or self-reprogramming" [118, p. 240]. And yet, granting these machines *any* capacity for change and evolution opens the door to unforeseen and potentially catastrophic outcomes.

Taking a somewhat different view, Freitas and Merkle later discussed the possibility of designing safe maker-replicator machines that are inherently *incapable* of undergoing evolution; they offered suggestions for how this might be achieved by "human-in-the-loop" approaches where we retain the ability to regulate the control architecture or supply of raw materials to the machines [119, p. 199]. They concluded by asserting that "[a]rtificial kinematic self-replicating systems which are not inherently safe should not be designed or constructed, and indeed should be legally prohibited."

Although NASA did not take their 1980 project forward, work on physical self-replicating systems for space exploration and exploitation has continued in various forms. A good review of developments in this area up to the early 2000s can be found in [119], and a review of more recent work is given in [212]. We highlight a few of these projects here just to give a flavour of recent developments.

A number of researchers have proposed the use of 3D printing as a practical means by which maker-replicator mining and manufacturing machines might be developed on the Moon. For example, Philip Metzger and colleagues' proposal, published in 2013, features an evo-maker-replicator approach that begins with a subreplicating system, remotely operated from Earth, and "evolves toward full self-sustainability ... via an in situ technology spiral" [206, p. 18]. The envisaged system would employ 3D printer-based manufacturing along with a range of other technologies. Metzger *et al.* argue that the development of such systems is now economically feasible because of the discovery of lunar polar ice, meaning that the Moon "has every element needed for healthy industry" [206, p. 18]. Echoing the dreams of Bernal and others before them, Metzger and colleagues suggest that their proposal would allow the production of material and energy resources that can be transported back to Earth, the terraforming of Mars, the establishment of space colonies in the solar system and, eventually, the colonisation of other solar systems [206, p. 28].

With similar goals to those of Metzger and colleagues, work by Alex Ellery addresses the challenge of designing self-replicators built only from materials available on the Moon [104, 105]. Ellery's approach is also based upon 3D printers, but with a particular focus upon what he regards as a key hurdle: the 3D printing of motors. In addition, he outlines approaches to solving other key aspects of a self-replicating machine, including printable electronics and sensors, and the chemical processing of raw materials. Ellery concludes that "[a]lthough there are many problems with which to contend, there appear to be no fundamental hurdles" [104, p. 325].

Elsewhere, Will Langford and colleagues have recently proposed an approach to reduce the complexity of physical self-replicators by using a small set of just thirteen basic part-types [183]. The part-types are categorised into four functional groups: structural, flexural, electronics and actuation. This work calls to mind Zuse's earlier proposal of simplfying the realisation of self-replication by using a modular design built from standardised parts [100].

A number of researchers have suggested biologically-based techniques for industrial activities in space. These include Lynn Rothschild and colleague's proposal for what they call *myco-architecture*, which uses bioengineered fungi to generate surface structures that could be grown in situ on other planets [254]. Another example is the recent *BioRock* experiment on the International Space Station, which studied the feasibility of using biomining (the use of microorganisms to extract valuable materials from ores) in microgravity environments [191]. If these kinds of technologies prove viable, it is easy to envisage how they could be incorporated as part of a bio-technological hybrid self-replicating system for space applications.

At a smaller scale, progress is being made towards the goal of molecular-level self-assembly and self-replication in the form of wetware and nanobot systems. Reviews covering various different flavours of this work can be found in [119, pp. 89–144, 201–217], [62, pp. 66–71], [243], [27], [95] and [314]. In addition to technical progress in these areas, there has also been much debate of the potential dangers of this work (e.g. [93, 23]). In 2000, the USA-based think tank the *Foresight Institute* published a set of guidelines for the development of nanotechnology, which particularly focused on replicator technology [113]. The guidelines recommended against the development of designs that could withstand mutation or undergo evolution. We look at more broad-ranging efforts to develop guidelines for the responsible development of advanced AI systems next.

6.4 Addressing the Risks Associated with Self-Replicators

In recent years there have been increasingly well-organised and multinational efforts to consider risks associated with the development of advanced AI technology.

Several governments (including the US [215], the European Parliament [106] and the UK [258]) have commissioned reports on the future of AI in order to develop appropriate policies in this area. The UK report noted that "the verification and validation of autonomous systems was 'extremely challenging' since they were increasingly designed to learn, adapt and self-improve during their deployment" [258, p. 16]. In developing the report for the European Parliament, a study for the Committee on Legal Affairs noted that "the self-replication of robots, and especially nanorobots, might prove difficult to control and potentially dangerous for humanity and the environment, thus requiring strict external control of research activities" [217, p. 11]. In a subsequent report for the Commission on Civil Law Rules on Robotics, the Committee on Civil Liberties, Justice and Home Affairs stated that "robotics and artificial intelligence, especially those with built-in autonomy, including the … possibility of self-learning or even evolving to self-modify, should be subject to robust conceptual laws or principles" [81, p. 36].

At the same time, several new institutes have been established to address these kinds of issues. A forerunner in this area is the *Foresight Institute*, mentioned in the previous section. Other more recent examples include the *Future of Life Institute* (Cambridge MA, USA), the *Future of Humanity Institute* (Oxford, UK), the *Centre for the Study of Existential Risk* (Cambridge, UK) and the *Machine Intelligence Research Institute* (Berkeley CA, USA). One example of their activities is the development (by the Future of Life Institute) of what has become known as the *Asilomar AI Principles*[15] to govern the safe, ethical development of powerful AI systems. At the time of writing, over 3,800 AI researchers and other endorsers have signed up to support these principles.[16] Principle number 22 states: "AI systems designed to recursively self-improve or self-replicate in a manner that could lead to rapidly increasing quality or quantity must be subject to strict safety and control measures." We will return to the discussion of risk management in self-replicator research in Chap. 7.

*   *   *

---

[15] https://futureoflife.org/ai-principles.

[16] https://futureoflife.org/principles-signatories/

We have now covered the full history of the idea of self-replicator technology: from the initial inklings of the notion of self-reproducing machines, first conceived of as standard-replicators in the seventeenth century (Chap. 2); followed by the additional idea born in the nineteenth century that machines might not only be able to reproduce but also to evolve—evo-replicators (Chaps. 3–4); up to the first serious theoretical treatments of the subject, the crystallisation of the new idea of maker-replicators and the first implementations of self-reproducing machines in the mid-twentieth century (Chap. 5); and ending with an overview of more recent developments up to the present day (this chapter). Turning to the final chapter, we will now offer some of our own thoughts on what has been achieved, the various goals that have driven this research, technical issues that remain unresolved and prospects for future developments.

Chapter 7
The Next Evolution: Reflection and Outlook

Having reached the end of our review, we now take a step back to assess the implications of the work we have described, the issues that remain unresolved, and the likely directions of future research. Later in the chapter we consider technical details and practical problems relating to implementations of self-replicators, and conclude with a discussion of what we consider to be the most likely directions for future developments. But first we consider the narratives and future worlds imagined by the earliest commentators.

7.1 Narratives of Self-Replicators

As demonstrated in the preceding chapters, the early history of thought about self-reproducing and evolving machines unveils a diverse array of hopes and fears. These contributions show that current debates about the implications of AI and ALife for the future development of humankind are actually a continuation of a conversation that has been in progress for many centuries. In this section we look at the main recurring themes that are apparent in the early works of scientific, philosophical and fictional literature. We focus in particular on the nineteenth century writing of Butler (Sect. 3.1), Marshall (Sect. 3.2) and Eliot (Sect. 3.3), the early twentieth century literary work of Forster (Sect. 4.1.1) and Čapek (Sect. 4.1.2), the early pulp sci-fi work by Wright, Campbell, Manning, Williams and Dick (Sect. 4.1.3), and Bernal's early scientific speculations (Sect. 4.2.1).

7.1.1 Takeover by Intelligent Machines

Perhaps the most prominent theme apparent in these works is the fear that machines might evolve to a level where they displace humankind as the dominant intelligent species. While some writers proposed more positive, co-operative alliances

© Springer Nature Switzerland AG 2020
T. Taylor, A. Dorin, *Rise of the Self-Replicators*,
https://doi.org/10.1007/978-3-030-48234-3_7

between humans and machines—including Butler, Marshall, Wright, Campbell and Bernal—none was fully convinced by this outcome, and all discussed less desirable possibilities elsewhere in their writings.[1]

The idea that we ourselves are creating our own successors can be seen in the work of Butler, Eliot, Čapek, Wright and Campbell. Some saw this not as a development to be feared but rather as a way in which the reach of humankind might be extended beyond the extinction of our species; examples include Čapek, Campbell (in *The Last Evolution*) and Williams.

Most saw the evolution of increasingly intelligent machines as an inevitable process. In the work reviewed in Chaps. 3–4, only Čapek engages significantly with the idea that humans might exert some control over the robots' reproduction. Less optimistically, Butler and Bernal thought this could likely only be achieved by humans forsaking the development of technology altogether.

The idea of *self-repairing* machines is present in the work of Eliot, Forster, Campbell (in *The Machine*) and Bernal, and this is indeed a theme in current evolutionary robotics research.[2] In contrast, we are unaware of any serious scientific investigation of the idea of *self-designing* machines, which appears in the sci-fi work of Wright, Campbell and Dick—the closest we get to it is in work on Lamarckian evolution, such as that of Richard Laing described in Sect. 6.1.[3] These sci-fi authors portray self-design as a route by which the pace of machine evolution can accelerate through a process of self-reinforcement; these works, and Butler's and Marshall's before them, strongly foreshadow current interest in the ideas of *superintelligence* and the *technological singularity*.[4]

7.1.2 Implications for Human Evolution

Beyond the idea that machines might become the dominant intelligent species, the reviewed works have explored a number of potential implications of self-reproducing machines for the future direction of human evolution.

In *Erewhon* Butler envisaged that humans might become weaker and physically degenerate due to reduced evolutionary selection pressure brought about by all-caring machines. Eliot and Forster foresaw a similar outcome. In contrast, an alternative outcome explored by Butler (in *Lucubratio Ebria*) and Bernal is that human abilities might become significantly *enhanced* by the incorporation of increasingly sophisticated cyborg technology.

---

[1] Marshall is a possible exception, although his goal was to propose a model of biological learning and intelligent behaviour rather than to predict the future of humankind.

[2] Examples include [28] and [71].

[3] Despite the lack of significant scientific developments in this area, the topic of self-designing machines is nevertheless a recurring theme in recent discussions about the future of AI. We say more about this in Sect. 7.1.4.

[4] We previously mentioned this in relation to the work of Butler (Sect. 3.1) and Marshall (Sect. 3.2). See in particular our comments in footnote 8, p. 20.

Several authors emphasised that humans and machines are engaged in a *co-evolutionary* process. In *Lucubratio Ebria* Butler suggested that this closely coupled evolution of humans and machines might increase our physical and mental capabilities. In particular, he proposed that intelligent machines might change the environment in which humans develop and evolve, thereby influencing our own evolutionary path and intertwining it with that of the machines; this idea foreshadows the modern concept of biological *niche construction* [219]. In *The Last Evolution* Campbell envisaged a positive outcome of this co-evolution, with human creativity working in harmony with machine logic and infallibility. Butler in *Erewhon*, however, was more dubious of the process, conjuring an image of machines as parasites benefiting from the unwitting assistance of humans in driving their evolution.

Beyond the discussion of evo-replicators, these early works also explored potential applications of standard-replicators. In particular, various authors envisaged these as a technology to allow humankind to explore and colonise other planets. The properties of self-repair and multiplication by self-reproduction are seen as essential for attempts to traverse the immense distances of interstellar—or even intergalactic—missions. Bernal's vision is of self-repairing and self-reproducing living environments to allow multiple generations of humans to survive such journeys. Williams, and Dick (in *Autofac*), have our robot successors making the journey in place of us. More recently, Tegmark (Sect. 6.1) suggested that an advanced AI might make the journey by itself but then rebuild the human race from manufactured DNA once it arrives at its destination.

7.1.3 Implications for Human Society

In addition to imagining consequences for human evolution, these authors also envisaged how human society and the lives of individuals might be affected by the existence of superintelligent machines.

The prospect of humans becoming mere servants to machines was raised by Butler (in *Darwin Among The Machines*), Wright and Manning. However, Butler suggests that this might not necessarily be a detrimental development—the machines would likely take good care of us, at least for as long as they still rely upon humans for performing functions relating to their maintenance and reproduction.

Many of the works explore how humans might spend their time in a world where all of their basic needs are taken care of by beneficent machines. In Forster's work, humans engage in the exchange of ideas and academic learning (mostly about the history of the world before the all-nurturing Machine existed). Similarly, Bernal suggests that we would be free to pursue science and also other areas of uniquely human activity including art and religion. Individuals in Campbell's *The Machine* are chiefly occupied with playing physical games and pursuing matters of the heart. They also develop an unhealthy reverence to the Machine as a god, to the extent that the Machine ultimately decides to leave that planet so that the humans can learn to live independently again.

Likewise, Butler (in *Erewhon*) and Bernal discuss the possibility that humans might separate from machines at some point in the future, although in their works, in contrast to Campbell's, this is a decision made by the humans rather than the machines. Bernal also considers the possibility that the human species might ultimately diverge into two, with one group pursuing the path of technological co-evolution, and the other rejecting technology and searching for a simpler and more satisfying existence more at one with nature.

7.1.4 The Narratives in Context

In surveying the futures envisaged by these early thinkers, we should be mindful of the potential for a dystopian bias in their works—the vast majority of which were written by young, white men [251].[5] Indeed, Max Tegmark has recently summarised a much broader range of alternatives for how the future relationship between humans and advanced AI might unfold, covering the whole utopian/dystopian spectrum [284]. It is certainly true that the large-scale mechanical self-reproducing machines envisaged by these early authors have not yet been realised. Nevertheless, as outlined in Chap. 6, research continues on the development of standard-replicators, evo-replicators and maker-replicators, in hardware and in software. Sustained thought, discussion and planning for a future shared with self-replicator technology is therefore essential.

In Chap. 1 we identified three major steps in the intellectual development of the field. It is instructive to consider how the context and assumptions of each of these steps have influenced the work described, and how alternative perspectives at each step might suggest different avenues of research.

The first step grew out of the idea that animals could be viewed as machines and vice versa (Sect. 2.1). This perspective will suggest very different kinds of self-reproducing machine depending on one's conception of the design of organisms. There were many different views on this topic in the seventeenth and eighteenth centuries (for a discussion of these, see, e.g. [250, ch. 3], [117], [94]). If, for example, one took the view of the eminent eighteenth century French physician Théophile de Bordeu, of the living body as a decentralised being akin to a swarm of bees ([123, pp. 138–139], [210, p. 56])—or, indeed, any of the subsequent views of organisms as self-organising systems, from Kant to Maturana and Varela [307]—one might arrive at a very different design for a self-reproducing machine than that instantiated in von Neumann's cellular model.

Rather than von Neumann's complex monolithic design, a "swarm-like" self-reproducing system might comprise a factory of thousands or millions of machines that achieve production closure, material closure and collective reproduction as a

---

[5] For a more general discussion of the role of cultural context in the portrayal of fictional robots, see [274].

whole (see Sect. 7.3.1 for further discussion of closure).[6] Indeed, the idea of the collective self-reproduction of a diverse group of machines was raised by Butler in *Erewhon* (Sect. 3.1) and was implicit in Čapek's play *R.U.R.* (Sect. 4.1.2). The idea was central to Konrad Zuse's concept of a "self-reproducing workshop" (Sect. 5.4.2) and also to some of the more recent proposals for space exploration and exploitation discussed in Sect. 6.3. However, few of the other recent software or hardware implementations mentioned in Chap. 6 have employed significantly decentralised designs.

Furthermore, there are other aspects of the design of self-reproducing machines that might be influenced by one's conception of the essential, relevant or typical traits of organisms—of what kind of thing a living organism *is*. The apparently self-generative nature of embryonic development has been a central topic of debate for biologists, physicians and philosophers from Aristotle to modern times ([216], [253], [250, ch. 8]). Von Neumann's self-reproducing automata in his cellular model build offspring by constructing a full "adult" copy of themselves as directed by the genetic information recorded on the information tape. In contrast, multicellular biological organisms pass on genetic information which enables their embryonic offspring to "build themselves"—and, in so doing, they allow for the development of the final form of the organism to be influenced epigenetically by the environment in which they find themselves. Few of the studies reviewed here have touched upon this topic, although it was raised as an issue by Dyson and was also discussed in the NASA study report [118, p. 199] (Sect. 6.3).[7]

As identified in Chap. 1, a vital component of the second major step of the intellectual development of the field was the acceptance of the idea that animals had evolved. Von Neumann's theoretical work, and the early experimental work on evo-replicators by Barricelli, Penrose and Jacobson, discussed in Chap. 5, adopted an essentially modern neo-Darwinist perspective. That is, the primary mechanism by which improvements could appear in these systems was by fortuitous mutations of the genetic information passed from parent to offspring.[8]

However, when designing self-reproducing machines, we have free rein to equip them with alternative mechanisms for transmitting information from one generation to the next, beyond genetic inheritance. We could, for example, equip the machines with the ability to engage in inter-generational learning and cultural transmission like human societies.

---

[6] Von Neumann's theoretical work on the logic of self-reproduction did not commit to any particular design, but the cellular model design has become particularly associated with his work as it was the only practical example that he produced before his death.

[7] There has been a small amount of work on this topic (e.g. [38, 39]) but much remains to be studied, especially in relation to utilising physical and self-organisational properties of the environment to influence and assist the development of the offspring.

[8] In Barricelli's studies (Sect. 5.2.1), he also observed the crossing of genetic material from one symbioorganism to another, which might be interpreted as genetic recombination or horizontal gene transfer, but these processes still fall within the neo-Darwinist picture.

More radically, we might also implement mechanisms that are completely unavailable to any biological species.[9] For example, if we have a particular goal in mind, we could apply directed mutations in a machine's genetic information to induce specific changes in its offspring. Similarly, the direct transmission to offspring of characteristics acquired during an individual's lifetime (Lamarckian evolution) is rejected as a mechanism for biological evolution by neo-Darwinism,[10] but it might nevertheless be possible, and even useful, for machine evolution. For example, we might equip a parent machine with the ability to directly copy its "brain state" (the state of its control systems after a lifetime of learning about its environment) directly into its offspring's brain.[11] As discussed in Sect. 6.1, there have been some limited explorations of the evolutionary potential of Lamarckian self-replicators. At the same time, various authors have questioned the reliability of Lamarckian reproduction architectures, specifically those implemented by means of a machine actively inspecting its own body (see, e.g., [4, pp. 211–214]). More research is required to really understand how the performance of these kinds of systems compares to standard neo-Darwinian designs.

Even more radically, a parent machine might create more advanced offspring by intentionally designing an improved form itself rather than relying upon genetic mutations or the cultural transmission of information. This notion of *self-designing* machines was present in some of the early sci-fi stories discussed in Sect. 4.1.3, but we are unaware of any serious scientific investigation of the idea. Despite the lack of tangible progress in this area, the potential of self-designing machines to develop advanced levels of intelligence, and to follow goals that are not necessarily aligned with our own, are very much topics of concern in current debates about the risks associated with the development of AI. In particular, it has been cautioned that a machine that can design a better version of itself could lead to a succession of ever more intelligent machines, each one an improvement on its predecessor, in a process of *recursive self-improvement* [29, p. 35]—a kind of supercharged evolutionary process.

As previously mentioned, a good discussion of the full range of possible outcomes of this kind of technology and their implications for humankind, spanning the complete spectrum from utopias to dystopias and various intermediate outcomes, has recently been provided by Max Tegmark [284, ch. 5]. The development of AI capable of recursive self-improvement is covered by the *Asilomar AI Principles* (Sect. 6.4); these were formulated at a meeting of some of the world's leading AI researchers in 2017, and they include a policy promoting strict safety and control measures for AI systems designed to recursively self-improve.

---

[9] The idea that we might be able to implement novel mechanisms to improve the efficiency of the evolutionary process in searching for specific goals in the context of AI is certainly not a new one. For example, it was discussed by Alan Turing in his seminal work *Computing Machinery and Intelligence*, published in 1950 [295, p. 456] (we mentioned Turing's work in Sect. 5.5).

[10] Note, however, that there is currently a renewed interest in the importance of some forms of transmission of acquired information in biological evolution [149, 181].

[11] Such a "Lamarckian" system was indeed discussed in the NASA report [118, p. 244].

Having considered the context in which these discussions and research unfolded, in the following sections we look at differences in the technical approaches adopted in the implementations of self-replicator technology described in Chaps. 5–6. We also highlight some of the technical issues that remain to be solved in this work, and offer suggestions of which particular lines of research are most likely to succeed in the short-term and the long-term future. In order to do that, it is helpful to first take another look at the various goals and purposes that different researchers have in mind when pursuing this work.

7.2 Purpose and Goals of Research on Self-Replicators

Throughout this book we have made the distinction between three different flavours of self-replicator. As introduced in Sect. 1.3, work on *standard-replicators* covers the basic design requirements and potential applications of machines that can faithfully produce copies of themselves; work on *evo-replicators* embraces the evolutionary potential of self-reproducing machines as a route to the automatic generation of complex AI; and work on *maker-replicators* emphasises the manufacturing possibilities of self-replicating universal constructors—many of those working in this area actively seek to *avoid* the possibility of evolution that might lead to unanticipated behaviours.

We can also make an orthogonal distinction between the *reasons* people have for pursuing this research. We can broadly categorise the projects described in our review as having either *scientific*, *commercial* or *sociological* goals as their driving forces.[12]

Scientific goals include contributing to our understanding of the origins of life and elucidating the general design of living organisms. Of the work we have reviewed, Barricelli (Sect. 5.2.1), Penrose (Sect. 5.3.1) and Jacobson (Sect. 5.3.2) were primarily interested in the former goal, whereas the latter was a component of von Neumann's interest in the topic (Sect. 5.1.1). Work towards these goals tends to focus primarily on evo-replicators (or in some cases, including von Neumann's work, on evo-maker-replicators).

The obvious commercial reason for pursuing research on maker-replicators is the potential to totally transform the economics of the production of goods, with the prospect of an exponentially increasing and theoretically unlimited yield from a fixed initial production cost.[13] This was a central component of Moore's discussion

---

[12] These are not necessarily exclusive categories. We might also add *philosophical* goals in some cases. We have not included *engineering* goals because the question of why one is trying to engineer a self-reproducing system would ultimately fall into one of the other categories specified.

[13] Of course, the yield would in practice be limited by the availability of resources. Any closed environment would therefore impose a ceiling on the population size of machines that it could support. This kind of restriction would be lessened in cases where machines could expand into new environments, such as travelling to other planets.

(Sect. 5.4.1), and it was discussed in more detail in later work by Dyson and in the NASA study (Sect. 6.3).

Commercial goals for evo-replicators include the evolution of artificial intelligence in a variety of settings. While some people view evolutionary ALife as a path to AGI (Sect. 6.1), others pursue it for different commercial reasons. One example is using evolution to generate rich virtual worlds populated by whole ecosystems of virtual organisms of different kinds; here, the focus is not on achieving human-level—or even human-like—intelligence, but on reproducing biological evolution's capacity to generate a wild diversity of interacting species of varying levels of complexity [278].

There are two major sociological reasons for studying self-reproducing systems that have emerged from our review and that are also evident in the discussion on narratives in the previous section. The first is the view that the evolution of technology is already an unstoppable process, and that self-reproducing machines may either be an inevitable component of humankind's future on Earth or may indeed displace us to become the dominant species. This was a central concern in Butler's work (Sect. 3.1) and also a common theme in sci-fi stories (Sect. 4.1.3). The second reason is that self-reproducing machines could be a means by which humans—or their technological offspring—might eventually colonise other solar systems and other galaxies. This was a core part of Bernal's investigation (Sect. 4.2.1), and it has been the focus of some of the more recent studies described in Sect. 6.3.

Bearing these different goals in mind, and also the underlying context and assumptions behind the work we have described, we now delve into a more detailed discussion of the contrasting approaches to the design and implementation of self-reproducing machines in the work we have reviewed. As we'll see, there is a strong connection between the goal of the research and the design approach adopted.

7.3 The Process of Self-Reproduction

As described in Sects. 5.2–5.3, the first implementations of self-reproducing systems, such as Barricelli's computational symbioorganisms and Penrose's physical blocks, were simple compositions of a small number of elementary units. These stand in great contrast to von Neumann's complex designs (Sect. 5.1.1). What are we to make of the contrast between these seemingly vastly different approaches?

In Chap. 1, we stated that no system was truly *self*-reproducing, but that the process is always the result of an interaction between a structure to be copied and the environment in which it exists. This observation has been emphasised by almost every author who has considered the technicalities of the process in detail, including most of those discussed in Chap. 5 such as von Neumann, Penrose, Jacobson and Ashby, and in Zuse's later work too (Sect. 6.3). The minimal level of complexity required in the design of a self-reproducing machine depends upon the environment in which it operates, and the extent to which the machine can utilise processes and features of the environment to aid its reproduction; the more the machine can "offload" the process of reproduction to the environment by relying upon the laws of physics[14] to do the job for it, the simpler the machine can be. On the other hand, the more the process of self-reproduction is explicitly controlled by the machine itself (thereby requiring a more complex machine), the wider the variety of different environments in which it might potentially be able to reproduce.

Answers to questions about the desired complexity of the elementary units of a self-reproducing machine, the features of a suitable environment in which it is to operate, and the appropriate relationship between the machine and its environment, depend on the researcher's goals.[15] In the previous section we already discussed the different reasons people have had for studying self-reproducing machines, and their associated goals.

One general observation apparent in the implementations reviewed in Chaps. 5–6 is that those focused on maker-replicators tend to be much more complex than those focused on evo-replicators. To examine this observation in more detail, in the following sections we discuss the differences between maker-replicator designs and evo-replicator designs. Maker-replicators, on the one hand, are exemplified by von Neumann's "top-down" design approach; that is, the starting point of his work was a theory-driven design of a complex machine (a universal constructor). Using this overall design as a guide, with much effort and ingenuity he then developed the low-level design details of a working implementation (his cellular model). This standard "engineering" approach to the problem resulted in a monolithic architecture of hundreds of thousands of parts. Evo-replicators, on the other hand, are exemplified by the "bottom-up" design approaches employed by Penrose, Barricelli and Jacobson; their designs comprised only a few relatively simple parts that, their designers anticipated, would evolve increased complexity over time.

7.3.1 Maker-Replicators: The Top-Down Approach

As described in Sect. 5.1.1, von Neumann was interested in producing machines that could perform arbitrary tasks of vast complexity. He used self-reproduction as a means to this end, realising that his objective could be achieved by designing a machine that could build another machine more complicated than itself. His goal required that the machines could perform other tasks in addition to reproduction, and that the complexity of the additional tasks could increase from parent to offspring. In other words, von Neumann's goal was to build not just a maker-replicator but an

---

[14] In physical systems, and also in the kind of "fully embedded" computational dynamical systems considered by von Neumann and by Barricelli, all action is ultimately determined by general laws of dynamics that act upon objects in the system. In the real world, these are the laws of physics and chemistry. In the following discussion, we use the term "laws of physics" to refer to these general laws of dynamics in any system, either physical or computational.

[15] See [276] and [277] for further discussion of this issue in the context of creativity in computational evolutionary systems.

evo-maker-replicator, and it is for these reasons that his architecture features *both* the capacity for universal construction *and* for evolvability.

The self-reproducing automata of von Neumann's cellular model were embedded in their environment; that is, they were made of the same elementary parts as the rest of the environment and were subject to the same dynamics or laws of physics. Von Neumann designed the model this way because the ability of the automata to operate upon the same "stuff" from which they were themselves made, and thereby to construct new automata themselves, was fundamental to the problem. However, the type of environment provided by the cellular model was very different to the physical environment experienced by biological organisms. It was a discrete, digital space lacking basic concepts from the physical world such as the conservation of matter or notions of energy or force. Of course, von Neumann deliberately set aside such issues in order to focus upon the logical issues involved in self-reproduction and the evolution of complexity.

For reproduction in a physical environment, von Neumann's architecture would need to be extended to deal with processes such as the collection, storage and deployment of material resources and energy.[16] Once these are included, it is likely that the machine would need to be able to withstand perturbations, maintain its organisation and self-repair.[17] Several authors have pointed out that real-world self-reproducing machines would also have to deal with clearing up dead parts and recycling parts if the environment is not to become clogged with waste (e.g. [151, p. 262], [96, p. 198], [118, p. 239]).

Even more fundamentally, when moving from the idealized space of von Neumann's cellular model to a physical implementation of a complex maker-replicator, issues relating to *closure* become substantially more challenging. We can separate these issues into two categories, those relating to *production closure* and those relating to *material closure*. The property of production closure is satisfied if every component of the self-reproducing machine can be constructed by the machine itself. The property of material closure is satisfied if the machine is able to collect from within its operating environment all of the raw materials required to build its offspring.[18]

Von Neumann's architecture for a self-reproducing machine provides a high-level design of one possible approach to achieving production closure, although it gives little specific guidance for how such a machine might be constructed in practice. His cellular model, and the follow-up studies by others mentioned in Sects. 6.1–6.2, provide example implementations in software, but these designs do not translate easily

---

[16] As mentioned in Sect. 5.1.1, von Neumann planned to return to some of these issues later [303, p. 82] but did not reach that stage before his early death.

[17] As discussed by Moore, it is possible that if the machine reproduced fast enough, then a certain level of failure could be tolerated, hence reducing the need for self-maintenance [208, p. 121]. However, if we wish to allow for the evolution of arbitrarily complex machines with potentially much longer net reproduction times, they would likely have to engage in self-maintenance at some stage.

[18] We might also add *energy closure* to this list—the property of the machine being able to obtain from its operating environment all of the energy required for its operation and reproduction. Alternatively, this could be viewed as an aspect of material closure.

to physical realizations where much more attention is required to considerations of materials, energetics and so on. Recent developments in 3D printing (Sect. 6.3) are moving in the direction of production closure, but the fact remains that this is still an unsolved problem for physical machines in the general case.

Production closure is what Arbib referred to as the "fixed point problem of components" (Sect. 6.1); he, and von Neumann before him, thought there would be some minimum level of complexity of machine that was able to achieve production closure. Thinking in terms of manufacturing machines made out of parts drawn from a relatively small list of basic types (e.g. sensors, motors, structural, computational, cutting, joining, etc.), von Neumann argued that "[t]here is a minimum number of parts below which complication is degenerative, in the sense that if one automaton makes another the second is less complex than the first, but above which it is possible for an automaton to construct other automata of equal or higher complexity" [303, p. 80]. Zuse's concept of a self-replicator's Rahmen (Sect. 6.3) is useful here, in reminding us that the threshold complexity required for production closure of a self-replicator will depend upon the complexity of the external facilities that it requires to sustain its activity. A small number of recent publications have reported advances in the theory of production closure (e.g. [156]), but it remains a core issue to be tackled in future work.

While von Neumann's work addressed at least the high-level logical aspects of production closure, it completely ignored issues relating to material closure. In his cellular model, the self-reproducing machine could generate new parts out of thin air when constructing its offspring. More recently, those working on maker-replicator designs for space applications have paid the most attention to this problem; examples include the 1980 NASA study and the work of Metzger and colleagues mentioned in Sect. 6.3. Nevertheless, the construction of a physical maker-replicator with full material closure remains a distant dream.

Zuse's idea of simplifying the design of a maker-replicator by employing a modular approach using standardised parts (Sect. 6.3) would presumably help to alleviate the problems associated with both types of closure. However, he did not complete a full design for a machine of this kind, nor has this been achieved in any subsequent work on physical maker-replicators.

If and when these issues of production closure and material closure in physical maker-replicators are resolved, solutions would still be required for the other problems mentioned relating to energetics, self-repair and dealing with waste products. It is theoretically possible that a human designer could develop a much more complicated version of von Neumann's self-reproducing machine which included all of these features. However, the collective experience of roboticists and AI researchers in the sixty years since von Neumann's death suggests that it is easier to design machines that can cope with unknown real-world environments by allowing them to learn and adapt, either by lifetime learning or by evolution. For real world applications, the possibility of a human designer foreseeing all possible situations and equipping the machine to deal with them is simply not a viable alternative.

One potential solution to this problem would be to design a much simpler replicating machine to place in the environment. The aim would then be to have it evolve towards the capacities of von Neumann's architecture. This approach might alleviate the need for a complex human-engineered machine designed from first principles. Instead, natural selection would test and pass (or fail) each aspect of the machine's design in the context of its environment.

To fully achieve von Neumann's vision, we might therefore have to (at least partially) tackle the "origins problem" (Sect. 5.1.1) that he had originally intended to set aside. This would greatly expand the breadth of problems to be addressed. We would like to ensure that the process eventually arrived upon a von Neumann-like architecture with universal construction capabilities, along with the additional capacities mentioned above for dealing with the physical realities of materials, energetics and so on—and preferably did so via a reasonably efficient route. This would present us with the challenge of how to guide evolution in the desired direction. As daunting as this seems, we would at least have biology to guide us, as this is precisely the challenge that biological evolution faced, and conquered, during the early stages of the development of terrestrial life.

However, endowing physical self-reproducing machines with the capacity to evolve is a strategy with many potential risks for our species, for our environment, and for life in general. As we have seen already, there are explicit cautions against developing these kinds of systems in the Foresight Institute guidelines (Sect. 6.3) and in the Asilomar AI Principles and other initiatives described in Sect. 6.4.

A somewhat similar proposal that relies less on the wide-ranging evolutionary potential of the self-reproducing system itself is Metzger and colleagues' idea of developing a fully self-reproducing system from an initially subreplicating version by an "in situ technology spiral" (Sect. 6.3). As we saw in Chaps. 5–6, various authors have focused more generally on the design of maker-replicators with less emphasis on the capacity for evolution (or in many cases with the desire to actively avoid any such capacity). Examples include the early speculations of Moore (Sect. 5.4.1) and some of the more recent studies on physical implementations by NASA and others (Sect. 6.3). The focus of many of these projects is on controllability and robust operation, meaning that they likely have less need to adapt and evolve than von Neumann's design. These kinds of architectures seem more likely than von Neumann's to be developed into practical physical implementations in the near- to mid-future.

7.3.2 Evo-Replicators: The Bottom-Up Approach

As described above, the first implementations of evo-replicator systems, such as Barricelli's computational symbioorganisms and Penrose's physical blocks, involve radically simpler designs than those employed in the maker-replicator studies of von Neumann and others.

The self-reproducing entities in these systems are aggregates built from only a handful of basic types of unit, and the basic units are assumed to exist in plentiful supply within the system's operating environment. This bottom-up approach of creating self-reproducing aggregates out of linear chains of simple parts significantly reduces the closure issues faced by more complex maker-replicators; but the price paid for this is the self-replicators having a greatly reduced behavioural repertoire—they are no longer capable of universal construction. The major question facing the bottom-up approach is therefore, is it possible for these simple self-reproducing aggregates to eventually evolve much more complex behaviour, and if so, how?

In the biological world, life has evolved from very simple beginnings to modern organisms that instantiate something like von Neumann's architecture as part of their design.[19] There are many open questions about how the genetic architecture evolved to its modern state, such as (1) what were the architectures of the original and intermediate stages of biological life? and (2) what properties of the architectures and the environment ensured that each stage had enough evolutionary potential to eventually bring forth the next stage? These are active research questions in studies of the origins of life, and results from that area will doubtless influence future work on designing evo-replicator machines.

A key question when designing self-replicator systems is, what is the appropriate level at which to start? Designing a mechanical equivalent to the hypothesised conditions of the origins of life could provide our system with the most unrestricted evolutionary potential, but we might have to wait a very long time for any complex or useful behaviour to emerge from it.[20] Some intermediate level between a primordial soup and a full implementation of von Neumann's architecture would be a more practical starting point. However, the appropriate starting point is heavily dependent on whether a project's focus is on maker-replicators or evo-replicators, and on the specific goals of each individual research programme.

It is useful to consider the simpler designs for self-reproduction studied by Penrose and Barricelli—how do the architectures of these systems influence their evolutionary potential?[21] Penrose's most complicated models (Sect. 5.3.1) allowed chains of arbitrary length (and therefore carrying an arbitrary amount of information) to be reproduced. But for the information to become evolutionarily relevant, it must have some effect on the chain's ability to reproduce. While Penrose discussed this issue

---

[19] We are referring here to the fundamental aspects of life's genetic architecture such as the translation of genetic information to determine an organism's form and activity, and the copying of the genetic information during reproduction. These aspects, which were central to von Neumann's reasoning about evolvability, are present in all modern organisms—not just in complex eukaryotic organisms such as ourselves but in bacteria and archaea as well.

[20] Various estimates of time spans for the artificial evolution of intelligent species have been proposed in the literature, including those of L. L. Whyte in the 1920s mentioned in Sect. 4.2.1 (footnote 11, p. 38).

[21] In Jacobson's implementation of self-reproduction in his model railway system (Sect. 5.3.2), the information driving the machine's operation was not explicitly copied from parent to offspring but assumed to be present in one of the elementary units. While he did discuss how the design might be significantly extended to allow for the explicit copying of this information, he did not implement such a system. Evolution in his implemented system would therefore be considerably restricted compared to the designs of Penrose or Barricelli, because it would rely upon fortuitous mutations of the elementary units themselves, whereas novel patterns in the other systems could emerge simply by recombining elementary units in different ways. Hence, we regard Jacobson's design as relatively impoverished compared to the others in this respect, and do not discuss it further here.

[234, pp. 112–114], Barricelli actually experimented with the idea by allowing his symbioorganisms to encode strategies for playing games that would determine their success at competing against neighbouring symbioorganisms for space (Sect. 5.2.1).

However, Barricelli's approach was deficient in that his symbioorganisms, unlike Penrose's linear chain replicators, could not carry arbitrary information. Only very specific configurations could be viable self-reproducers because they were collectively autocatalytic organisations; that is, their constituent elements all had to be placed in particular positions relative to each other in order for the structure as a whole to reproduce. Furthermore, while the approach provided the symbioorganisms with some phenotypic "toy bricks" to play with, the system was designed with a simple fixed mechanism for translating a symbioorganism's configuration into a game playing strategy. *Tac Tix* was (literally) the only game in town, and if and when a symbioorganism mastered it, there was no other avenue along which it might improve itself.

7.3.3 Top-Down and Bottom-Up Approaches Compared

In contrasting von Neumann's architecture for an evo-replicator (specifically, an evo-maker-replicator) with examples of trivial self-reproduction, author William Poundstone remarked that "[t]he important thing was that the self-reproducing know-how reside in the aggregate machine rather than in any of the raw materials" [239, p. 131]. As we stated earlier, this issue of where the "self-reproducing know-how" resides was discussed by von Neumann (Sect. 5.1.1), Penrose (Sect. 5.3.1), Jacobson (Sect. 5.3.2) and Zuse (Sect. 6.3), among others. The progressively more explicit specification of the method of reproduction by the machine itself is potentially self-reinforcing, as the more the know-how resides in the information stored in the machine rather than in the laws of physics, the more subject to mutation and evolution the process will be; this could eventually lead to the emergence of more sophisticated and complex forms of reproduction.

Although some of the evo-replicator systems designed or proposed by Penrose and Barricelli *did* allow the machines to explicitly carry information of potential relevance to their chances of reproduction,[22] a significant difference between their designs and von Neumann's is that in the latter, the information passed from parent to offspring was processed by an interpreter that was *itself part of the machine*, and was therefore also described on the machine's information tape. Hence, the information on the tape could be expressed in an arbitrary language defined by the interpreter. This opens up the possibility that the language in which genetic information is expressed could itself evolve, becoming progressively more efficient at expressing how to construct complex machines. In contrast, in Penrose's and Barricelli's systems the information was processed according to a fixed language of

---

[22] Specifically, Barricelli's *Tac Tix*-playing symbioorganisms (Sect. 5.2.1), and Penrose's discussion of how machines based on his most complex designs might perform tasks dependent upon their configuration (Sect. 5.3.1).

interpretation, which we could regard as being part of the laws of physics of the system.

However, in terms of the capacity for this language to evolve further, von Neumann's design was deficient for the same reasons as the architecture in general—it was a complex human-designed architecture that was introduced into the environment without having been through the filter of natural selection from simple beginnings to ensure its robustness and evolvability in its environment.[23]

With regard to evo-replicator design, Pask had already suggested in the cybernetics literature of the early 1960s that the evolution of the genetic language was an important issue (Sect. 5.5). It also became a core question in Barricelli's later work (Sect. 5.2.1, especially [18]). More recently, Howard Pattee has explored the topic in detail, in the context of both hardware and software implementations of artificial life (e.g. [230]).

Beyond a straight comparison of "top-down" and "bottom-up" approaches, we should also remember that other designs are possible too. As discussed in Sect. 7.1, ideas such as collectively self-reproducing factories of machines, Lamarckian evolution systems and intentionally self-designing systems all suggest alternative architectures. Further research is required to properly understand the strengths and weaknesses of each of these approaches and to ascertain which might be the most appropriate solution for any given project.

7.3.4 Drive for Ongoing Evolution

Even if we managed to address all of the issues outlined above, evo-replicator developers would still be faced with the question of how to provide the *drive* for ongoing evolution of the system. Pask (Sect. 5.5) suggested that in an ecosystem of self-reproducing machines, such drive would come from co-evolutionary interactions between the machines. Barricelli's symbioorganisms had already provided an example of this process in action (Sect. 5.2.1). In *Lucubratio Ebria*, Butler envisaged a co-evolutionary process not between machines and other machines, but between machines and humans (Sect. 3.1). Bernal considered a mixture of the two processes, with his artificial planets (globes) competing for natural resources and also directed by the desires of their colonists (Sect. 4.2.1). The question of how to build systems that possess continual evolutionary activity leading to the ongoing discovery of new adaptations and innovations is the central focus of current research on *open-ended evolution* by the Artificial Life community (Sect. 6.1).

For those following von Neumann's grand goals of creating self-reproducing general manufacturing machines which also have the ability to evolve (i.e. evo-

---

[23] Von Neumann acknowledged this point, stating that mutations to the interpreting machine would generally result in unviable offspring [303, p. 86] (but see [137] and [22] for some recent investigations into the evolvability of the architecture). Although his architecture implemented a sophisticated *epistemic cut* between organism and environment [230], there is little evidence that it has the capacity for significant further evolution.

maker-replicators), many questions remain to be answered if this is to become a safe and commercial technology of practical benefit to us, rather than an avenue by which we might unwittingly create our own successors.

One key outstanding question is how to provide a drive towards performing *specific tasks*. We would need to understand how to reliably direct the evolution of the machines' behaviour to fulfil specific human needs—while avoiding unwanted or harmful side effects—in addition to honing their own needs for survival and reproduction. Our experience of the devastation caused by invasive biological species could pale into insignificance compared to the havoc that evo-maker-replicators (or evo-replicators in general) might wreak. Biological invasive species at least have a shared evolutionary history that unites all terrestrial life at the level of basic biochemistry. Physical evo-replicators would lack this shared ancestry—they would be alien species in the very strongest sense. Even before they had evolved any particularly complex or intelligent behaviour, the very simplest physical evo-replicators might in themselves represent an existential threat to humankind. If they evolved and speciated much faster than their biological counterparts, they could generate their own parallel ecosystem which might rapidly dissolve the indigenous one (*our* ecosystem) by depriving it of its essential resources including matter, energy and simply the space in which to live. As we have shown in the preceding chapters, the dangers of self-replicators developing undesirable behaviours unaligned with our own needs has been a common theme in the early literature. This is an example of what has become known in current discussions about AI safety as the *value alignment problem* [255].

On the other hand, those working with some applications of evo-replicators actively seek to create systems where replicators *can* develop their own goals and desires, beyond those set for them by their human designers. This is particularly true in scientific research on understanding how the autonomous generation of goals has arisen in the biological world, and also in the development of virtual ecosystems for entertainment purposes. Open-ended evolution could be a route by which the evolution of goals, desires and purposiveness is achieved. The filter of natural selection applied to a population of evo-replicators ensures that only those individuals whose constitutions (i.e. their organisation and behaviour) are best adapted for survival and reproduction persist. This leads to the evolution of replicators whose constitutions are strongly aligned with their goals. In other words, natural selection results in a situation where *the existence, design and behaviour of an evo-replicator can all be explained in terms of how they promote the replicator's survival and reproduction.* In the biological world, this is the process by which organisms have attained the ability to act according to their own rules of behaviour rather than merely being passively acted upon by the laws of physics [231]. Furthermore, in biology we see that different species have evolved a dizzying variety of *instrumental* goals on top of their *final* goals—that is, we see that many different strategies and ways of living have emerged to achieve the same underlying goals of survival and reproduction.

Taking a lesson from nature, the open-ended evolution of evo-replicators by natural selection is therefore a potential route by which AIs could develop true agency—the ability to develop and act according to their own constitution-aligned goals. This

potential of evolution to engender purposiveness and agency is not currently a major focus of research in open-ended evolution,[24] but we expect that to change in the
coming years.

To delve much deeper into these issues would take us too far away from the
historical focus of this book. Suffice it to say, there are plenty of suggestions in the
origins of life literature about how genetic systems might evolve from very simple
beginnings to the level of complexity observed in modern biological organisms.[25]
Furthermore, there is a growing literature on mechanisms by which innovations
arise in evolutionary processes, which will also be of significant relevance to future
work.[26]

7.4 Looking Forward

As our review has shown, the notion of self-replicator technology has captured the
imagination of scientists, writers and the public alike for a remarkably long time.
The roots of the idea can be traced back to early comparisons between animals and
machines in the seventeenth and eighteenth centuries (Chap. 2), conjuring the first
inklings of standard-replicator machines. Speculation about the future potential of
the technology, and its implications for our own species, blossomed in the nineteenth century in the wake of the British Industrial Revolution and the publication
of Darwin's theory of evolution by natural selection (Chap. 3). These developments
heralded the emergence of the idea of evo-replicator machines. Self-replicator technology was a recurring theme in science fiction stories and other literary works in
the early twentieth century (Sect. 4.1), and received the first rigorous scientific treatment by John von Neumann in the 1940s (Sect. 5.1.1). The first realisations in both
digital and physical forms soon followed in the 1950s (Sects. 5.2–5.3), accompanied
by a distinct line of research focused on maker-replicator machines. As we outlined
in Chap. 6, more recent decades have seen continued progress in all areas, with
further research and development of standard-replicator, evo-replicator and maker-
replicator technology both in physical form and in software implementations.

However, notwithstanding the quotes shown in Chap. 1 and the recent developments described in Chap. 6, the idea has fallen out of the media spotlight. Despite
the steady progress described in Chap. 6, there have been no really major recent
breakthroughs in the area, unlike in other areas of AI and machine learning that
currently command so much attention from the mass media.

It is true that no one has yet succeeded in building a large-scale physical self-
reproducing machine of the kind envisaged by von Neumann or NASA. While von
Neumann's work showed that it was theoretically possible to build a self-replicator

---

[24] Although there are some hints at it elsewhere in the recent literature (e.g. [284, pp. 253–255],
[187]).

[25] Many of these are reviewed in [171] and [84].

[26] Examples of work on evolutionary innovations from a biological perspective include [142],
[200], [304] and [305]. Examples from an ALife perspective include [226], [279] and [172].

(indeed, one featuring both universal construction and evolvability—an evo-maker-replicator) without any logical paradox or infinite regress of description, various critical practical issues remained unaddressed. Not least among these were the questions of how the self-replicator might ensure a continual supply of energy and raw materials. Despite some recent progress in these areas, such as the latest work on maker-replicators for space systems mentioned in Sect. 6.3, these questions still represent key hurdles for researchers working on physical self-replicator technology.

The continual supply of energy and raw materials are less daunting issues when we consider molecular-level self-replicating systems (Sect. 6.3). It is for this reason, combined with potentially lower development costs, that we believe significant near-term progress in physical self-reproduction is most likely to occur in these kinds of systems (i.e. wetware and nanobot maker-replicators). With molecular-level systems, as with physical self-reproducing systems at any scale, development of this technology must be accompanied by careful consideration of potential risks including the possibility of environmental havoc caused by an out-of-control self-replication process. If these systems have the potential to evolve, then the hazards are further amplified. Examples of current efforts to mitigate, control and govern these risks include those described in Sect. 6.4. A great deal more effort will be required as this technology continues to develop, addressing possible risks in all media in which self-replicators could be developed, in hardware, software and molecular-level systems.

Over a slightly longer time frame of several decades, and funding permitting, work on large-scale physical self-replicators in the form of maker-replicators is likely to become a more significant enterprise. We consider the most likely applications of this technology to be in space exploitation and exploration. This would represent the realisation of ideas first put forward by J. D. Bernal nearly one hundred years ago and also envisaged by Konrad Zuse, Freeman Dyson and others (Sect. 6.3). The NASA study of 1980 represents the most significant effort in this area to date, but new technologies and scientific discoveries have provided extra impetus to this field in the last decade (Sect. 6.3). Of these new technologies, recent initial explorations of biologically-based techniques for off-Earth mining and construction might eventually provide the easiest route to developing large-scale physical self-replicators by creating the possibility of a bio-technological hybrid approach.

Notwithstanding these developments in physical self-replicator technology, the most active area of current research is undoubtedly in software systems (Sect. 6.2). In contrast to research on physical systems, the majority of contemporary work on software self-replicators focuses upon the evolutionary potential of self-reproducing agents—that is, evo-replicators rather than maker-replicators. Rather than trying to restrict the possibility of self-replicators to evolve, this work actively seeks to understand the biological world's capacity for continual inventiveness, and to create software systems that exhibit similarly open-ended evolutionary dynamics. Some view these kinds of evolutionary artificial life systems as a promising route to achieving human- or superhuman-level artificial general intelligence (AGI). Related to this, evolution by natural selection can furnish an AI with purposiveness and true agency—the ability to act according to their own goals and desires (Sect. 7.3.4). More mundanely, but no less importantly, work on software-based self-replicator technology could also become a useful test bed for understanding the effectiveness of measures proposed to curb the evolutionary capacity of physical self-replicator systems. In addition to these mid- to long-term applications, the technology also has commercial applications in the short-term, such as providing a means of populating open virtual worlds with a rich diversity of lifeforms.

It is tempting to think that work on software-based self-replicator technology is, in itself, a much safer pursuit than its hardware counterparts. Yet there is no room for complacency here, because the boundaries between the virtual and physical worlds are inexorably dissolving. Examples such as the malicious Stuxnet computer worm, which is believed to have caused targeted real-world damage to Iran's nuclear-enrichment facilities [176], give some indication of the potential dangers.

Within the last decade we have become accustomed to headline-grabbing discussions of grave dangers connected with the development of AGI, superintelligence and the hypothesised technological singularity. In the near-term at least, it is the potential of malicious or out-of-control software self-replicators to cause disruption and damage, whether targeted or unintended, both in the virtual world and in the real world, that represents the most pressing risk of this technology. Recent years have seen the emergence of various initiatives aimed at understanding the risks associated with the advent of advanced AI including self-replicator technology, and at providing guidelines for the responsible development of these systems (Sect. 6.4). Nevertheless, the history of computer security suggests that we can expect an ongoing battle between those who develop harmful software evo-replicators (either intentionally or through ignorance or negligence) and those who seek to protect their online and real-world systems from potential damage by such systems.

Before work commenced on the first implementations of software and hardware self-replicators in the 1950s (Chap. 5), the concerns of earlier commentators in the nineteenth and early twentieth centuries were mostly about the possibility of large-scale physical evo-replicators and the consequences of this technology for the future of humankind. However, in light of the challenges and complexity involved in their design, the likely costs versus short-term benefits of their development and the risks involved in their operation, we do not envisage this particular kind of self-replicator technology as representing a danger in the near-term. Of the various kinds of systems we have discussed, including evo-replicators and maker-replicators in software and in hardware, these large-scale physical evo-replicators are the least likely to be developed any time soon.

Nevertheless, as the work we have reviewed in the preceding chapters demonstrates, the goal of building large-scale evo-replicators is a persistent idea that has occupied the minds of forward thinkers from the publication of *On the Origin of Species* over one hundred and sixty years ago to the present day. The hurdles that must be overcome to implement this kind of system are immense, but they do not appear to be completely insurmountable.

Farmer and Belin (Sect. 6.1 and quoted on p. 1) suggest that the impact of physical evo-replicator technology "on humanity and the biosphere could be enormous, larger than the industrial revolution, nuclear weapons, or environmental pollution" [109, p. 815]. As envisaged by various authors we have discussed, this technology could be a means by which humankind assures its long-term survival across deep time and space by providing a route by which we might colonise the universe, or by evo-replicators becoming our worthy successors. On the other hand, it also has the potential to wreak havoc in the environment, to disrupt the biosphere, to develop its own goals unaligned with our own and to wipe us out in the process, and, through these processes, to ultimately extinguish the light of consciousness in the universe.

In this, as with all other forms of self-replicator technology, whether it is ultimately beneficial or detrimental to us depends upon how well we understand the issues at stake, and upon how that understanding enables us to properly manage its development. A thorough understanding of these issues should be based upon a sound appreciation of the history of the ideas involved. It is our hope that the review and discussion we have set out in the preceding chapters represents a helpful starting point in this endeavour.

References

[1] M. Amos. *Genesis Machines*. Atlantic Books Ltd, 2006.

[2] P. Anderson. Epilogue. In *Analog: Science Fact and Science Fiction*, pages 112–158. Condé Nast, New York, NY, Mar. 1962.

[3] P. J. Angeline. A historical perspective on the evolution of executable structures. *Fundamenta Informaticae*, 35(1-4):179–195, 1998.

[4] M. A. Arbib. Self-reproducing automata—some implications for theoretical biology. In C. Waddington, editor, *Towards a Theoretical Biology*, volume 2, pages 204–226. Edinburgh University Press, 1969.

[5] M. A. Arbib. The likelihood of the evolution of communicating intelligences on other planets. *Interstellar Communication: Scientific Perspectives*, pages 59–78, 1974.

[6] W. B. Arthur. *The Nature of Technology: What it is and how it evolves*. Free Press, New York, NY, 2009.

[7] P. M. Asaro. From mechanisms of adaptation to intelligence amplifiers: The philosophy of W. Ross Ashby. In P. Husbands, O. Holland, and M. Wheeler, editors, *The Mechanical Mind in History*, chapter 7, pages 149–184. MIT Press, Cambridge, MA, 2008.

[8] W. R. Ashby. The self-reproducing system. In C. A. Muses and W. S. McCulloch, editors, *Aspects of the Theory of Artificial Intelligence (The Proceedings of the First International Symposium on Biosimulation, Locarno, June 29–July 5, 1960)*, chapter 2, pages 9–18. Springer, New York, NY, 1962.

[9] I. Asimov, P. S. Warrick, and M. H. Greenberg, editors. *Machines That Think: the best science fiction stories about robots and computers*. Holt, Rinehart and Winston, New York, NY, 1983.

[10] W. Aspray and A. Burks, editors. *Papers of John von Neumann on Computing and Computer Theory*. MIT Press, Cambridge, MA, 1987.

[11] W. Banzhaf and L. Yamamoto. *Artificial Chemistries*. MIT Press, Cambridge, MA, 2015.

[12] N. A. Barricelli. Esempi numerici di processi di evoluzione. *Methodos*, 6:45–68, 1954.

© Springer Nature Switzerland AG 2020
T. Taylor, A. Dorin, *Rise of the Self-Replicators*,
https://doi.org/10.1007/978-3-030-48234-3

[13] N. A. Barricelli. Symbiogenetic evolution processes realized by artificial methods. *Methodos*, 9(35–36):143–181, 1957.

[14] N. A. Barricelli. Theory testing by numerical evolution experiments. Technical report, Vanderbilt University, Nashville, TN, May 1959.

[15] N. A. Barricelli. Numerical testing of evolution theories. Part I. Theroetical introduction and basic tests. *Acta Biotheoretica*, XVI(1/2):69–98, 1962.

[16] N. A. Barricelli. Numerical testing of evolution theories. Part II. Preliminary tests of performance. Symbiogenesis and terrestrial life. *Acta Biotheoretica*, XVI(3/4):99–126, 1963.

[17] N. A. Barricelli. Numerical testing of evolution theories. *Journal of Statistical Computation and Simulation*, 1:97–127, 1972.

[18] N. A. Barricelli. Suggestions for the starting of numeric evolution processes to evolve symbioorganisms capable of developing a language and technology of their own. *Theoretic Papers*, 6(6):119–146, 1987. (A publication of the Blindern Theoretic Research Team, University of Oslo).

[19] J. D. Barrow and F. J. Tipler. *The Anthropic Cosmological Principle*. Oxford University Press, Oxford, 1986.

[20] N. W. Bartlett, M. T. Tolley, J. T. Overvelde, J. C. Weaver, B. Mosadegh, K. Bertoldi, G. M. Whitesides, and R. J. Wood. A 3D-printed, functionally graded soft robot powered by combustion. *Science*, 349(6244):161–165, 2015.

[21] G. Basalla. *The Evolution of Technology*. Cambridge University Press, Cambridge, U.K., 1988.

[22] D. Baugh and B. McMullin. Evolution of G-P mapping in a von Neumann self-reproducer within Tierra. In *Advances in Artificial Life, ECAL 2013: Proceedings of the Twelfth European Conference on the Synthesis and Simulation of Living Systems*, pages 210–217, Cambridge, MA, 2013. MIT Press.

[23] R. Baum. Nanotechnology: Drexler and Smalley make the case for and against 'molecular assemblers'. *Chemical & Engineering News*, 81(48):37–42, 2003.

[24] G. Beauchamp. Technology in the dystopian novel. *Modern Fiction Studies*, 32(1):53–63, 1986.

[25] J. D. Bernal. *The Physical Basis of Life*. Routledge and Kegan Paul, London, UK, 1951.

[26] J. D. Bernal. *The World, the Flesh and the Devil: An enquiry into the future of the three enemies of the rational soul*. Jonathan Cape, London, 2nd edition, 1970. (Second edition with a new Foreword by the Author. First edition published in 1929).

[27] A. J. Bissette and S. P. Fletcher. Mechanisms of autocatalysis. *Angewandte Chemie International Edition*, 52(49):12800–12826, 2013.

[28] J. Bongard, V. Zykov, and H. Lipson. Resilient machines through continuous self-modeling. *Science*, 314(5802):1118–1121, 2006.

[29] N. Bostrom. *Superintelligence: Paths, dangers, strategies*. Oxford University Press, Oxford, 2014.

[30] A. Bowyer. Self-reproducing machines and manufacturing processes. In Y. Bar-Cohen, editor, *Biomimetics: Nature-Based Innovation*, page 361–376. CRC Press, 2011.

[31] P. Brantlinger. *Taming Cannibals: Race and the Victorians*. Cornell University Press, Ithaca, NY, 2011.

[32] J. Breivik. Self-organization of template-replicating polymers and the spontaneous rise of genetic information. *Entropy*, 3(4):273–279, 2001.

[33] S. Brenner. *My Life in Science*. BioMed Central, 2001. As told to Lewis Wolpert. Editors: Errol C. Friedberg and Eleanor Lawrence.

[34] S. P. Brilmyer. "The Natural History of My Inward Self": Sensing character in George Eliot's *Impressions of Theophrastus Such*. *PMLA*, 129(1):35–51, 2014.

[35] L. Brodbeck, S. Hauser, and F. Iida. Morphological evolution of physical robots through model-free phenotype development. *PloS one*, 10(6):e0128444, 2015.

[36] L. Brodbeck and F. Iida. An extendible reconfigurable robot based on hot melt adhesives. *Autonomous Robots*, 39(1):87–100, 2015.

[37] A. Brown. *J. D. Bernal: The Sage of Science*. Oxford University Press, 2005.

[38] W. R. Buckley. Computational ontogeny. *Biological Theory*, 3(1):3–6, 2008.

[39] W. R. Buckley. Computational ontogeny. In A. Rosa, A. Dourado, K. Madani, J. Filipe, and J. Kacprzyk, editors, *Proceedings of the 4th International Joint Conference on Computational Intelligence (IJCCI 2012)*, pages 116–121, Setúbal, Portugal, 2012. SciTePress.

[40] S. Bullock. Charles Babbage and the emergence of automated reason. In P. Husbands, O. Holland, and M. Wheeler, editors, *The Mechanical Mind in History*, chapter 2, pages 19–39. MIT Press, Cambridge, MA, 2008.

[41] E. Bulwer-Lytton. *The Coming Race*. William Blackwood and Sons, Edinburgh, 1871.

[42] S. Butler. Darwin on the Origin of Species. *The Press*, 20 December 1862. https://paperspast.natlib.govt.nz/newspapers/press/1862/12/20/2.

[43] S. Butler. Darwin Among the Machines. *The Press*, 13 June 1863. https://paperspast.natlib.govt.nz/newspapers/press/1863/6/13/1.

[44] S. Butler. *A first year in Canterbury Settlement*. Longman, Green, Longman, Roberts and Green, 1863. Available online at http://www.gutenberg.org/ebooks/3235.

[45] S. Butler. Lucubratio Ebria. *The Press*, 29 July 1865. https://paperspast.natlib.govt.nz/newspapers/press/1865/7/29/2.

[46] S. Butler. The Mechanical Creation. In G. J. Holyoake, editor, *The Reasoner*. London, 1 July 1865.

[47] S. Butler. *Erewhon*. Trübner & Co., London, 1872. (Page numbers cited in text are from the Penguin Classics edition of 1985, edited with an Introduction by Peter Mudford).

[48] S. Butler. *The Note-Books of Samuel Butler*. A. C. Fifield, 1912. Edited by Henry Festing Jones. http://www.gutenberg.org/files/6173/6173-h/6173-h.htm.

[49] S. Butler. *Further Extracts from the Note-Books of Samuel Butler.* Jonathan Cape, London, 1934. Chosen and edited by A. T. Bartholomew.

[50] S. Butler and M. Butler. *The Correspondence of Samuel Butler with His Sister May.* University of California Press, Berkeley, CA, 1962. Edited with an Introduction by Daniel F. Howard.

[51] S. Butler and E. M. A. Savage. *Letters between Samuel Butler and Miss E. M. A. Savage 1871–1885.* Jonathan Cape, London, 1935.

[52] B. Cahill. Geneticist says life-like 'thing' could be created in test tube. *Montreal Gazette*, page 23, 25 August 1958.

[53] E. Calamy, editor. *The Works of the Rev. John Howe, M.A.: Memoirs of his Life*, volume II, chapter *The Principles and Oracles of God: Lecture IV*, pages 1059–1061. John P. Haven, New York, NY, 1838.

[54] J. W. Campbell. The Last Evolution. *Amazing Stories*, pages 414–421, Aug. 1932.

[55] J. W. Campbell. The Machine. *Astounding Stories*, pages 70–82, Feb. 1935. (Published under the pseudonym Don A. Stuart).

[56] K. Čapek. R.U.R. Rossum's Universal Robots; kolektivní drama v vstupní komedii a tech aktech, 1920. Online facsimile version of the 1920 first edition in Czech available at https://archive.org/details/rurrossumsuniver00apekuoft.

[57] K. Čapek. *Two Plays by Karel Čapek: R. U. R. (Rossum's Universal Robots) & The Robber.* Booksplendour Publishing, Brisbane, Australia, 2008. Translated and with an Introduction by Voyen Koreis.

[58] R. B. Carter. Descartes's bio-physics. In G. J. D. Moyal, editor, *René Descartes: Critical Assessments*, volume IV, chapter 103, pages 194–219. Routledge, London, 1991.

[59] J. Case. Periodicity in generations of automata. *Theory of Computing Systems*, 8(1):15–32, 1974.

[60] S. Cave, C. Craig, K. S. Dihal, S. Dillon, J. Montgomery, B. Singler, and L. Taylor. Portrayals and perceptions of AI and why they matter. Technical report, The Royal Society, Nov. 2018. DES5612 (doi:10.17863/CAM.34502).

[61] W. Churchill. Fifty years hence. *Strand Magazine*, Dec. 1931. Text available online at http://teachingamericanhistory.org/library/document/fifty-years-hence/. A slightly edited version appeared one month earlier in the Canadian magazine Maclean's (15 November 1931: https://archive.macleans.ca/issue/19311115).

[62] M. Ciofalo. Green grass, red blood, blueprint: Reflections on life, self-replication, and evolution. In J. A. Bryant, M. A. Atherton, and M. W. Collins, editors, *Design and Information in Biology: From Molecules to Systems*, volume 27 of *WIT Transactions on State of the Art in Science and Engineering*, chapter 3, pages 29–96. WIT Press, Southampton, UK, 2006.

[63] A. C. Clarke. *Greetings, Carbon-Based Bipeds! A vision of the 20th century as it happened (Collected Essays 1934–1998).* Voyager, London, 1999. (Edited by Ian T. MacAuley).

[64] J. Cohen. *Human Robots in Myth and Science.* George Allen & Unwin Ltd, London, 1966.

[65] K. K. Collins. G. H. Lewes revised: George Eliot and the moral sense. *Victorian Studies*, 21(4):463–492, 1978.

[66] S. Cook. Minds, machines and economic agents: Cambridge receptions of Boole and Babbage. *Studies in History and Philosophy of Science*, 36:331–350, 2005.

[67] J. B. Copeland and D. Proudfoot. Alan Turing's forgotten ideas in computer science. *Scientific American*, pages 99–103, 1999.

[68] B. Corry. Marshall, Alfred. In *International Encyclopedia of the Social Sciences*. Thomson Gale, 2008. http://www.encyclopedia.com/people/social-sciences-and-law/economics-biographies/alfred-marshall.

[69] J. Cottingham. A brute to the brutes? Descartes's treatment of animals. In G. J. D. Moyal, editor, *René Descartes: Critical Assessments*, volume IV, chapter 111, pages 323–331. Routledge, London, 1991.

[70] M. Črepinšek, S.-H. Liu, and M. Mernik. Exploration and exploitation in evolutionary algorithms: a survey. *ACM Computing Surveys (CSUR)*, 45(3):35, 2013.

[71] A. Cully, J. Clune, D. Tarapore, and J.-B. Mouret. Robots that can adapt like animals. *Nature*, 521(7553):503, 2015.

[72] C. Darwin. *On the Origin of Species by Means of Natural Selection, or the Preservation Of Favored Races in the Struggle for Life*. John Murray, London, 1859.

[73] E. Darwin. *Zoonomia: Or the Laws of Organic Life*. Joseph Johnson, London, 1794. (In two volumes).

[74] R. Dawkins. *The Selfish Gene*. Oxford University Press, Oxford, 1976.

[75] R. Dawkins. *The Extended Phenotype*. Oxford University Press, Oxford, 1982.

[76] B. L. B. de Fontenelle. *Oeuvres de Monsieur de Fontenelle, des Académies, Françoise, des Sciences, des Belles-Lettres, de Londres, de Nancy, de Berlin, & de Rome*, volume I, chapter *Galantes: Lettre XI—A Monsieur C...*, pages 321–323. Brunet, Paris, 1752.

[77] J. O. de La Mettrie. *L'Homme Machine*. 1747. (In "LaMettrie's L'Homme Machine: a Study in the Origins of an Idea", Aram Vartanian, Princeton University Press, 1960).

[78] D. J. de Solla Price. Automata and the origins of mechanism and mechanistic philosophy. *Technology and Culture*, 5(1):9–23, 1964.

[79] C. Dean. '55 'Origin of Life' Paper is Retracted. New York Times, 25 October 2007. http://www.nytimes.com/2007/10/25/science/25jacobson.html.

[80] L. del Ray. Though dreamers die. *Astounding Science Fiction*, pages 34–51, Feb. 1944.

[81] M. Delvaux. Report with recommendations to the commission on civil law rules on robotics. A8-0005/2017. European Parliament Committee on Legal Affairs, Jan. 2017. http://www.europarl.europa.eu/doceo/document/A-8-2017-0005_EN.html.

[82] P. J. Denning. The science of computing: The internet worm. *American Scientist*, 77(2):126–128, 1989. http://www.jstor.org/stable/27855650.

[83] D. Des Chene. *Spirits and Clocks: Machine and Organism in Descartes.* Cornell University Press, Ithaca, NJ, 2000.

[84] M. Di Giulio. The origin of the genetic code: theories and their relationships, a review. *Biosystems*, 80(2):175–184, 2005.

[85] P. K. Dick. Second variety. *Space Science Fiction*, pages 102–144, May 1953. Available at http://www.gutenberg.org/ebooks/32032.

[86] P. K. Dick. Autofac. *Galaxy Science Fiction*, pages 70–95, Nov. 1955.

[87] D. Diderot. Entretien entre M. d'Alembert et M. Diderot (Conversation between d'Alembert and Diderot). In *Le Rêve D'Alembert (D'Alembert's Dream)*. 1769. First published in von Grimm (ed.) *La Correspondance littéraire*, 1782; first published in its own right in 1830.

[88] B. Disraeli. *Coningsby: Or, The New Generation.* Henry Colburn, London, 1844. (Page numbers cited in text are from the Nonsuch Classics edition of 2007).

[89] A. Dneprov. Kraby idut po ostrovu. In *Znanie Sila*. Moscow, Russia, Nov. 1958.

[90] A. Dneprov. Crabs on the island. In *The Molecular Cafe: Science-Fiction Stories*. Mir, Moscow, Russia, 1968. English translation of the original 1958 Russian version.

[91] T. Doolittle. *The Young Man's Instructor and the Old Man's Remembrancer.* Thomas Parkhurst, London, 1673. (N.B. the author's surname on the book is spelt Doolittel, but the spelling Doolittle is now commonly used to refer to him and his work.).

[92] A. Dorin, K. Korb, and V. Grimm. Artificial-life ecosystems: What are they and what could they become? In S. Bullock, J. Noble, R. A. Watson, and M. A. Bedau, editors, *Proceedings of the Eleventh International Conference on Artificial Life*, pages 173–180, Cambridge, MA, 2008. MIT Press.

[93] K. E. Drexler. *Engines of Creation: The Coming Era of Nanotechnology.* Doubleday, New York, NY, 1986.

[94] F. Duchesneau. The organism-mechanism relationship: An issue in the Leibniz-Stahl controversy. In O. Nachtomy and J. E. H. Smith, editors, *The Life Sciences in Early Modern Philosophy*, chapter 6, pages 98–114. Oxford University Press, Oxford, 2014.

[95] H. Duim and S. Otto. Towards open-ended evolution in self-replicating molecular systems. *Beilstein Journal of Organic Chemistry*, 13(1):1189–1203, 2017.

[96] F. Dyson. *Disturbing the Universe.* Harper & Row, New York, NY, 1979.

[97] G. Dyson. *Darwin Among The Machines.* Addison-Wesley, 1997.

[98] G. Dyson. *Turing's Cathedral: The Origins of the Digital Universe.* 2012, New York, NY, Vintage Books.

[99] A. H. Eden, J. H. Moor, J. H. Søraker, and E. Steinhart, editors. *Singularity Hypotheses: A Scientific and Philosophical Assessment.* Springer, Berlin, 2013.

[100] N. Eibisch. Eine Maschine baut eine Maschine baut eine Maschine .... *Kultur & Technik (das Magazin aus dem Deutschen Museum)*, 1:48–

51, 2012. http://www.deutsches-museum.de/verlag/kultur-technik/archiv/36-jhrg-2012/.

[101] N. Eibisch. *Selbstreproduzierende Maschinen: Konrad Zuses Montagestraße SRS 72 und ihr Kontext*. Springer Vieweg, Wiesbaden, 2016.

[102] G. Eliot. *Impressions of Theophrastus Such: Essays and Leaves from a Note-Book*. William Blackwood & Sons, Edinburgh, 1879.

[103] G. Eliot. *Impressions of Theophrastus Such*. University of Iowa Press, Iowa City, 1994. This edition Edited and with Introduction by Nancy Henry. Originally published in 1879.

[104] A. Ellery. Are self-replicating machines feasible? *Journal of Spacecraft and Rockets*, 53(2):317–327, 2016.

[105] A. Ellery. Building physical self-replicating machines. In C. Knibbe, G. Beslon, D. Parsons, D. Misevic, J. Rouzaud-Cornabas, N. Bredèche, S. Hassas, O. Simonin, and H. Soula, editors, *Proceedings of the 14th European Conference on Artificial Life 2017 (ECAL 2017)*, pages 146–153, Cambridge, MA, 2017. MIT Press.

[106] European Parliament. Civil law rules on robotics. P8_TA(2017)0051. Texts Adopted (Thursday, 16 February 2017 – Strasbourg), Feb. 2017. http://www.europarl.europa.eu/sides/getDoc.do?type=TA&reference=P8-TA-2017-0051&language=EN&ring=A8-2017-0005.

[107] A. B. Evans, I. Istvan Csicsery-Ronay Jr, J. Gordon, V. Hollinger, R. Latham, and C. McGuirk, editors. *The Wesleyan Anthology of Science Fiction*. Wesleyan University Press, Middletown, CT, 2010.

[108] H. Fancher and M. Green. Blade Runner 2049. http://www.imdb.com/title/tt1856101/, 2017. Alcon Entertainment.

[109] J. Farmer and A. Belin. Artificial life: The coming evolution. In C. Langton, C. Taylor, J. Farmer, and S. Rasmussen, editors, *Artificial Life II*, volume X of *SFI Studies in the Sciences of Complexity*, Redwood City, CA, 1991. Addison-Wesley.

[110] H. Festing Jones. *Samuel Butler, Author of Erewhon (1835–1902): A Memoir*, volume 1. Macmillan & Co., London, 1919.

[111] D. B. Fogel. *Evolutionary Computation: The Fossil Record*. IEEE Press, Piscataway, NJ, 1998.

[112] D. B. Fogel. Nils Barricelli–artificial life, coevolution, self-adaptation. *IEEE Computational Intelligence Magazine*, 1(1):41–45, 2006.

[113] Foresight Institute. Foresight guidelines for responsible nanotechnology development. Current version published at https://foresight.org/about-nanotechnology/foresight-guidelines/.

[114] D. R. Forsdyke. Heredity as transmission of information: Butlerian 'intelligent design'. *Centaurus*, 48:133–148, 2006.

[115] E. M. Forster. The machine stops. In *The Oxford and Cambridge Review*. London, UK, Nov. 1909.

[116] E. M. Forster. A book that influenced me. In *Two Cheers for Democracy*. Edward Arnold, 1951.

[117] D. C. Fouke. Mechanical and "organical" models in seventeenth-century explanations of biological reproduction. *Science in Context*, 3(2):365–381, 1989.

[118] R. A. Freitas Jr and W. P. Gilbreath, editors. *Advanced Automation for Space Missions: Proceedings of the 1980 NASA/ASEE Summer Study*, NASA Conference Publication 2255, 1982. Available at https://ntrs.nasa.gov/search.jsp?R=19830007077.

[119] R. A. Freitas Jr and R. C. Merkle. *Kinematic Self-Replicating Machines*. Landes Bioscience, Georgetown, TX, 2004.

[120] A. R. Galloway. Creative evolution. *Cabinet Magazine*, 42:45–50, 2011. http://cabinetmagazine.org/.

[121] A. R. Galloway. The computational image of organization: Nils Aall Barricelli. *Grey Room*, 46:26–45, 2012.

[122] M. Gardner. Mathematical games: Concerning the game of nim and its mathematical analysis. *Scientific American*, pages 104–111, Feb. 1958.

[123] S. Gaukroger. *The Natural and the Human: Science and the Shaping of Modernity, 1739-1841*. Oxford University Press, Oxford, UK, 2016.

[124] D. J. Gillot. *Authority, Authorship, and Lamarckian Self-Fashioning in the Works of Samuel Butler*. PhD thesis, Birkbeck, University of London, 2013.

[125] S. B. Gissis and E. Jablonka. *Transformations of Lamarckism: From subtle fluids to molecular biology*. The Vienna Series in Theoretical Biology. MIT Press, Cambridge, MA, 2011.

[126] A. Gray. Darwin on the origin of species. *The Atlantic Monthly*, 6(33):109–116, July 1860.

[127] A. Gray. *Natural Selection not inconsistent with Natural Theology. A free examination of Darwin's Treatise on the Origin of Species, and of its American Reviewers*. Trübner & Co., London, 1861.

[128] J. Griesemer. The units of evolutionary transition. *Selection*, 1(1-3):67–80, 2001.

[129] S. Griffiths, D. Goldwater, and J. M. Jacobson. Self-replication from random parts. *Nature*, 437:636, 2005.

[130] R. Hackett. Meet the father of digital life. *Nautilus* magazine (web publication), June 2014. http://nautil.us/issue/14/mutation/meet-the-father-of-digital-life.

[131] G. S. Haight. *George Eliot: A Biography*. Oxford University Press, 1968.

[132] M. F. Hale, E. Buchanan, A. F. Winfield, J. Timmis, E. Hart, A. E. Eiben, M. Angus, F. Veenstra, W. Li, R. Woolley, M. De Carlo, and A. M. Tyrrell. The ARE robot fabricator: How to (re)produce robots that can evolve in the real world. In H. Fellermann, J. Bacardit, Ángel Goñi Moreno, and R. M. Füchslin, editors, *The 2019 Conference on Artificial Life*, pages 95–102, Cambridge, MA, 2019. MIT Press.

[133] A. W. Hall. Direct creation versus spontaneous generation and natural selection. *The Microcosm*, 8(11):161–166, Oct. 1891.

[134] Y. N. Harari. *Homo Deus: A Brief History of Tomorrow*. Harvill Secker, London, UK, 2015.

[135] O. Harman. *The Price of Altruism: George Price and the search for the origins of kindness*. The Bodley Head, London, UK, 2010.

[136] H. Harris. Lionel Sharples Penrose. 1898–1972. *Biographical Memoirs of Fellows of the Royal Society*, 19:521–561, 1973.

[137] T. Hasegawa and B. McMullin. Exploring the point-mutation space of a von Neumann self-reproducer within the Avida world. In *Advances in Artificial Life, ECAL 2013: Proceedings of the Twelfth European Conference on the Synthesis and Simulation of Living Systems*, pages 316–323, Cambridge, MA, 2013. MIT Press.

[138] L. J. Henkin. *Darwinism in the English Novel 1860–1910*. Corporate Press Inc., New York, 1940. (Page numbers cited in text are from the Russell & Russell edition of 1963).

[139] T. Hey and G. Pápay. *The Computing Universe: A Journey through the Revolution*. Cambridge University Press, Cambridge, UK, 2014.

[140] J. Hiller and H. Lipson. Automatic design and manufacture of soft robots. *IEEE Transactions on Robotics*, 28(2):457–466, 2012.

[141] T. Hobbes. *Leviathan: or, The Matter, Forme and Power of a Common Wealth Ecclesiasticall and Civil*. Andrew Crooke, London, 1651. (Page numbers cited in text are from the Wordsworth Classics of World Literature edition of 2014).

[142] M. E. Hochberg, P. A. Marquet, R. Boyd, and A. Wagner. Innovation: an emerging focus from cells to societies. *Philosophical Transactions of the Royal Society of London B: Biological Sciences*, 372(1735), 2017.

[143] G. M. Hodgson. The Mecca of Alfred Marshall. *The Economic Journal*, 103:406–415, Mar. 1993.

[144] D. Holloway. Innovation in science—the case of cybernetics in the Soviet Union. *Science Studies*, 4(4):299–337, 1974.

[145] K. C. Holmes. The life of a sage. *Nature*, 440:149–150, 2006.

[146] L. E. Holt. *Samuel Butler*. Twayne Publishers, revised edition, 1989.

[147] D. Howard, A. E. Eiben, D. F. Kennedy, J.-B. Mouret, P. Valencia, and D. Winkler. Evolving embodied intelligence from materials to machines. *Nature Machine Intelligence*, 1(1):12, 2019.

[148] F. Hoyle. *The Intelligent Universe: A New View of Creation and Evolution*. Holt, Rinehart and Winston, New York, NY, 1983.

[149] E. Jablonka and M. J. Lamb. *Evolution in Four Dimensions: Genetic, Epigenetic, Behavioral, and Symbolic Variation in the History of Life*. MIT Press, Cambridge, MA, 2005.

[150] H. Jacobson. Information, reproduction and the origin of life. *American Scientist*, 43(1):119–127, Jan. 1955.

[151] H. Jacobson. On models of reproduction. *American Scientist*, 46(3):255–284, Sept. 1958.

[152] H. Jacobson. No time like the present. *American Scientist*, 95(6):468, November-December 2007. (Letters to the Editors).

[153] B. C. Jantzen. *An Introduction to Design Arguments*. Cambridge Introductions to Philosophy. Cambridge University Press, Cambridge, U.K., 2014.

[154] R. D. Johnson and C. Holbrow. Space settlements: A design study. Technical report, NASA, Washington, D.C., NASA-SP-413 1977. (A report of the 1975 NASA Summer Study on Space Settlements. Available at https://ntrs.nasa.gov/search.jsp?R=19770014162).

[155] J. Johnston. *The Allure of Machinic Life: Cybernetics, Artificial Life, and the New AI.* MIT Press, Cambridge, MA, 2008.

[156] P. T. Kabamba, P. D. Owens, and A. G. Ulsoy. The von Neumann threshold of self-reproducing systems: theory and application. *Robotica*, 29(1):123–135, 2011.

[157] G. Kampis. *Self-Modifying Systems In Biology And Cognitive Science: A New Framework For Dynamics, Information And Complexity*, volume 6 of *IFSR International Series on Systems Science and Engineering*. Pergamon Press, Oxford, 1991.

[158] M. Kang. *Sublime Dreams Of Living Machines: The Automaton in the European Imagination.* Harvard University Press, Cambridge, MA, 2011.

[159] S. A. Kauffman. Autocatalytic sets of proteins. *Journal of theoretical biology*, 119(1):1–24, 1986.

[160] J. E. Kelleam. Rust. *Astounding Science Fiction*, pages 133–140, Oct. 1939.

[161] J. G. Kemeny. Man viewed as a machine. *Scientific American*, 192(4):58–67, 1955.

[162] Z. X. Khoo, J. E. M. Teoh, Y. Liu, C. K. Chua, S. Yang, J. An, K. F. Leong, and W. Y. Yeong. 3D printing of smart materials: A review on recent progresses in 4D printing. *Virtual and Physical Prototyping*, 10(3):103–122, 2015.

[163] D. King-Hele. Designing better steering for carriages (and cars); with a glance at other investions. In C. Smith and R. Arnott, editors, *The Genius of Erasmus Darwin*, chapter 14, pages 197–216. Ashgate, Aldershot, England, 2005.

[164] U. C. Knoepflmacher. *Religious Humanism and the Victorian Novel: George Eliot, Walter Pater and Samuel Butler.* Princeton University Press, Princeton, NJ, 1965.

[165] A. N. Kolmogorov. Automata and life (abstract of a paper read at the Methodological Seminar of the Faculty of Mechanics and Mathematics of Moscow State University of 5 April 1961). *Machine Translation and Applied Linguistics (Mashinnyy Perevod i Prikladnaya Lingvistika)*, 6:3–8, 1961. (In Russian).

[166] A. N. Kolmogorov. Automatons and living beings. In *Foreign Developments in Machine Translation and Information Processing*, volume 95, pages 1–4. U.S. Department of Commerce, Office of Technical Services, Joint Publications Research Service, Washington D.C., 1962. Report no. JPRS 13790 (CSO: 3901-D). (English Translation of six articles in Mashinnyy Perevod i Prikladnaya Lingvistika, No. 6, Moscow, 1961.).

[167] A. N. Kolmogorov. Life and thinking as special forms of the existence of matter. In G. M. Frank and A. M. Kuzin, editors, *On the Essence of Life*, pages 48–57. Science (Nauka), Moscow, 1964. (In Russian).

[168] A. N. Kolmogorov. Automata and life. In A. I. Berg, E. Kolman, and V. D. Pekelis, editors, *Cybernetics Expected and Cybernetics Unexpected (Kibernetika Ozhidaemaya i Kibernetika Neozhidannaya)*, pages 12–31. Science (Nauka), Moscow, 1968. (In Russian).

[169] A. N. Kolmogorov. Life and thinking as special forms of the existence of matter. In B. P. Konstantinov and V. D. Pekelis, editors, *Inhabited Space (Naselennyy Kosmos)*, pages 27–32. Science (Nauka), Moscow, 1972. (In Russian).

[170] A. N. Kolmogorov. Life and thinking as special forms of the existence of matter. In *Inhabited Space (Part One)*, pages 31–36. NASA Technical Translation, 1975. Report no. TT F-819. English Translation of Inhabited Space (Naselennyy Kosmos), Nauka Press, Moscow, 1974.

[171] E. V. Koonin and A. S. Novozhilov. Origin and evolution of the genetic code: the universal enigma. *IUBMB Life*, 61(2):99–111, 2009.

[172] K. B. Korb and A. Dorin. Evolution unbound: releasing the arrow of complexity. *Biology & Philosophy*, 26(3):317–338, 2011.

[173] B. M. Kozo-Polyanski. *Novyi printzip biologii. Ocherk teorii simbiogeneza.* Puchina, Leningrad, 1924. (In Russian).

[174] B. M. Kozo-Polyanski, V. Fet, and L. Margulis. *Symbiogenesis: a new principle of evolution.* Harvard University Press, Cambridge, MA, 2010. (Translated into English by V. Fet).

[175] R. Kurzweil. *The singularity is near: When humans transcend biology.* Penguin, 2005.

[176] D. Kushner. The real story of Stuxnet. *IEEE Spectrum*, 50(3):48–53, 2013.

[177] R. Laing. Some alternative reproductive strategies in artificial molecular machines. *Journal of Theoretical Biology*, 54(1):63–84, 1975.

[178] R. Laing. Automaton models of reproduction by self-inspection. *Journal of Theoretical Biology*, 66(3):437–456, 1977.

[179] R. Laing. Artificial organisms: History, problems, directions. In C. G. Langton, editor, *Artificial Life: The Proceedings of an Interdisciplinary Workshop on the Synthesis and Simulation of Living Systems (September 1987)*, volume VI of *Santa Fe Institute Studies in the Sciences of Complexity*, pages 49–61, Redmond City, CA, 1989. Addison-Wesley.

[180] R. A. Laing. *Automaton Self-Reference.* PhD thesis, State University of New York at Binghamton, 1977. Logic of Computers Group, Computer and Communication Sciences Department Technical Report No. 204.

[181] K. N. Laland, T. Uller, M. W. Feldman, K. Sterelny, G. B. Müller, A. Moczek, E. Jablonka, and J. Odling-Smee. The extended evolutionary synthesis: its structure, assumptions and predictions. *Proceedings of the Royal Society of London B: Biological Sciences*, 282(1813), 2015.

[182] B. Landon. Computers in science fiction. In J. Gunn, M. S. Barr, and M. Candelaria, editors, *Reading Science Fiction*, chapter 7. Palgrave Macmillan, Basingstoke, UK, 2008.

[183] W. Langford, A. Ghassaei, B. Jenett, and N. Gershenfeld. Hierarchical assembly of a self-replicating spacecraft. In *2017 IEEE Aerospace Conference*, pages 1–10, Piscataway, NJ, 2017. IEEE.

[184] C. G. Langton, editor. *Artificial Life: Proceedings of An Interdisciplinary Workshop On The Synthesis And Simulation Of Living Systems*, Santa Fe Institute Studies in the Sciences of Complexity, Boston, MA, 1989. Addison-Wesley Longman Publishing Co.

[185] J.-Y. Lee, J. An, and C. K. Chua. Fundamentals and applications of 3D printing for novel materials. *Applied Materials Today*, 7:120–133, 2017.

[186] S. Lem. *The Invincible*. Seabury Press, Inc., New York, NY, 1973. (Translated from the Polish version first published in 1964).

[187] S. R. Levin, T. W. Scott, H. S. Cooper, and S. A. West. Darwin's aliens. *International Journal of Astrobiology*, 18(1):1–9, 2019.

[188] S. Levy. *Artificial Life: The Quest for a New Creation*. Pantheon Books, New York, 1992.

[189] G. H. Lewes. *The Physical Basis of Mind*, volume 2 of *Problems of Life and Mind*. Trübner & Co., London, 1877.

[190] H. Lipson and J. B. Pollack. Automatic design and manufacture of robotic lifeforms. *Nature*, 406(6799):974–978, 2000.

[191] C.-M. Loudon, N. Nicholson, K. Finster, N. Leys, B. Byloos, R. Van Houdt, P. Rettberg, R. Moeller, F. M. Fuchs, R. Demets, J. Krause, M. Vukich, A. Mariani, and C. Cockell. Biorock: new experiments and hardware to investigate microbe–mineral interactions in space. *International Journal of Astrobiology*, 17(4):303–313, 2018.

[192] J. Lough, editor. *Diderot: Selected Philosophical Writings*. Cambridge University Press, 1953.

[193] G. Lucas and J. Hales. Star Wars: Episode II Attack of the Clones. http://www.imdb.com/title/tt0121765/quotes/qt0414984, 2002. Lucasfilm.

[194] R. MacCurdy, R. Katzschmann, Y. Kim, and D. Rus. Printable hydraulics: A method for fabricating robots by 3D co-printing solids and liquids. In *Robotics and Automation (ICRA), 2016 IEEE International Conference on*, pages 3878–3885. IEEE, 2016.

[195] J. Macmurray. *The Self as Agent: being the Gifford Lectures delivered in the University of Glasgow in 1953*. Faber and Faber, London, 1956. (Full text available at http://www.giffordlectures.org/books/self-agent).

[196] D. Mange, M. Sipper, A. Stauffer, and G. Tempesti. Towards robust integrated circuits: The embryonics approach. *Proceedings of the IEEE*, 88(4):516–543, 2000.

[197] L. Manning. The call of the mech-men. *Wonder Stories*, pages 366–385, Nov. 1933.

[198] A. Marshall. *The Principles of Economics*. Macmillan, London, 9th (variorum) edition, 1961. With annotations by C. W. Guillebaud.

[199] H. R. Maturana and F. J. Varela. *Autopoiesis and cognition: the realization of the living*. D. Reidel, Holland, 1972.

[200] J. Maynard Smith and E. Szathmáry. *The Major Transitions in Evolution*. W.H. Freeman, Oxford, 1995.

[201] A. Mayor. *Gods and Robots: Myths, Machines, and Ancient Dreams of Technology*. Princeton University Press, 2018.

[202] B. Mazlish. *The Fourth Discontinuity: The Co-Evolution of Humans and Machines*. Yale University Press, Newhaven, CT, 1993.

[203] B. McMullin. John von Neumann and the evolutionary growth of complexity: Looking backward, looking forward.... *Artificial Life*, 6(4):347–361, 2000.

[204] L. F. Menabrea. *Sketch of the Analytical Engine invented by Charles Babbage (with notes by the translator)*. Richard and John E. Taylor, London, 1843. (Translation and additional notes by Augusta Ada King, Countess of Lovelace).

[205] A. Mesoudi. Pursuing Darwin's curious parallel: Prospects for a science of cultural evolution. *PNAS*, 114(30):7853–7860, 2017.

[206] P. T. Metzger, A. Muscatello, R. P. Mueller, and J. Mantovani. Affordable, rapid bootstrapping of the space industry and solar system civilization. *Journal of Aerospace Engineering*, 26(1):18–29, 2013.

[207] M. Mitchell. *An Introduction To Genetic Algorithms*. MIT Press, Cambridge, MA, 1996.

[208] E. F. Moore. Artificial living plants. *Scientific American*, pages 118–126, Oct. 1956.

[209] H. Moravec. *Mind children: The future of robot and human intelligence*. Harvard University Press, Cambridge, MA, 1988.

[210] S. Moravia. From homme machine to homme sensible: Changing eighteenth-century models of man's image. *Journal of the History of Ideas*, 39(1):45–60, 1978.

[211] H. J. Morowitz. A model of reproduction. *American Scientist*, 47(2):261–263, June 1959.

[212] M. S. Moses and G. S. Chirikjian. Robotic self-replication. *Annual Review of Control, Robotics, and Autonomous Systems*, 3:5.1–5.24, 2020. (The version referred to in this book is the Review in Advance copy, first posted on the journal's website on October 21, 2019).

[213] E. Musk. Address at Tesla Annual Shareholder Meeting, May 2016. https://electrek.co/2016/06/01/elon-musk-machines-making-machines-rant-about-tesla-manufacturing/.

[214] O. Nachtomy and J. E. H. Smith. Introduction. In O. Nachtomy and J. E. H. Smith, editors, *The Life Sciences in Early Modern Philosophy*, chapter 1, pages 1–28. Oxford University Press, Oxford, 2014.

[215] National Science and Technology Council Committee on Technology. Preparing for the future of artificial intelligence. Executive Office of the President, United States of America, Oct. 2016. https://obamawhitehouse.archives.gov/sites/default/files/whitehouse_files/microsites/ostp/NSTC/preparing_for_the_future_of_ai.pdf.

[216] J. Needham and A. Hughes. *The History of Embryology*. Cambridge University Press, Cambridge, UK, 2nd edition, 1959.

[217] N. Nevejans.   European   civil   law   rules   in   robotics.   PE
      571.379. European   Parliament   Policy   Department   C: Citizens'
      Rights   and   Constitutional   Affairs,   Oct.   2016.   http://www.
      europarl.europa.eu/RegData/etudes/STUD/2016/571379/IPOL
      _STU(2016)571379_EN.pdf.

[218] B. Nieuwentyt.   *The Religious Philosopher: or, the Right Use of
      Contemplating the Works of the Creator*.   J. Senex and W. Taylor,
      London, 1718.   Translated by John Chamberlayne.   Available online
      at https://babel.hathitrust.org/cgi/pt?id=mdp.39015062240117. The watch-
      maker analogy appears on pp. lxv–xlvii.

[219] F. J. Odling-Smee, K. N. Laland, and M. W. Feldman. *Niche construction:
      the neglected process in evolution*. Princeton University Press, Princeton, NJ,
      2003.

[220] C. Ofria and C. O. Wilke. Avida: A Software Platform for Research in Com-
      putational Evolutionary Biology. *Artificial Life*, 10(2):191–229, Mar. 2004.

[221] S. J. Olson. Homogeneous cosmology with aggressively expanding civiliza-
      tions. *Classical and Quantum Gravity*, 32(21):215025, 2015.

[222] G. K. O'Neill. *The High Frontier: Human Colonies in Space*. Jonathan Cape,
      London, 1977.

[223] J. Ortega y Gasset. *Toward a Philosophy of History*, chapter 1: *The Sportive
      Origin of the State*. W. W. Norton & Co., New York, NY, 1941.

[224] J. Ortega y Gasset. *Obras Completas*, volume II, chapter *El origen deportivo
      del Estado* (originally published in 1924), pages 607–624. Revista de Occi-
      dente, Madrid, 1946.

[225] W. Ostwald. Machines and living creatures: Lifeless and living transformers
      of energy. *Scientific American Supplement*, 70(1803supp):55, July 1910.

[226] N. Packard, M. A. Bedau, A. Channon, T. Ikegami, S. Rasmussen, K. O.
      Stanley, and T. Taylor. An overview of open-ended evolution: Editorial intro-
      duction to the open-ended evolution II special issue. *Artificial Life*, 25(2):93–
      103, 2019.

[227] M. R. Page. *The Literary Imagination from Erasmus Darwin to H. G. Wells:
      Science, Evolution, and Ecology*. Ashgate Publishing Ltd., Farnham, UK,
      2012.

[228] W. Paley. *Natural Theology: or Evidences of the Existence and Attributes
      of the Deity, collected from the appearances of nature*. R. Faulder, London,
      1802. (Page numbers cited in text are from the Oxford University Press edi-
      tion of 2006, with an Introduction written by Matthew D. Eddy and David
      Knight).

[229] G. Pask. *An Approach to Cybernetics*. Hutchinson, London, 1961.

[230] H. H. Pattee.   Artificial life needs a real epistemology.   In F. Morán,
      A. Moreno, J. Merelo, and P. Chacón, editors, *Advances in Artificial Life:
      Third European Conference on Artificial Life*, Lecture Notes in Artificial In-
      telligence, pages 23–38, Berlin, 1995. Springer.

[231] H. H. Pattee. Evolving self-reference: Matter, symbols, and semantic closure.
      Communication and Cognition—Artificial Intelligence, 12(1–2):9–28, 1995.

[232] R. Pearl. *The Biology of Death*. J. B. Lippincott Co., Philadelphia, 1922.

[233] L. S. Penrose. Mechanics of self-reproduction. *Annals of Human Genetics*, 23(1):59–72, 1958.

[234] L. S. Penrose. Self-reproducing machines. *Scientific American*, pages 105–114, June 1959.

[235] L. S. Penrose. A theory of DNA replication. *Annals of Human Genetics*, 24:359–366, 1960.

[236] L. S. Penrose. On living matter and self-replication. In I. J. Good, editor, *The Scientist Speculates: An Anthology of Partly-Baked Ideas*, pages 258–271. Heinemann, 1962.

[237] L. S. Penrose and R. Penrose. A self-reproducing analogue. *Nature*, 4571:1183, June 1957.

[238] U. Pesavento. An implementation of von Neumann's self-reproducing machine. *Artificial Life*, 2(4):337–354, 1995.

[239] W. Poundstone. *The Recursive Universe: Cosmic Complexity and the Limits of Scientific Knowledge*. William Morrow, New York, NY, 1985.

[240] G. R. Price. The maker of machines. *Computers and Automation*, Oct. 1957.

[241] T. Raffaelli. The early philosophical writings of Alfred Marshall. Part I: Marshall's analysis of the human mind. *Research in the History of Economic Thought and Methodology*, Archival Supplement 4:55–93, 1994.

[242] T. Raffaelli. The early philosophical writings of Alfred Marshall. Part II: Marshall's papers. *Research in the History of Economic Thought and Methodology*, Archival Supplement 4:95–159, 1994.

[243] S. Rasmussen, M. A. Bedau, L. Chen, D. Deamer, D. C. Krakauer, N. H. Packard, and P. F. Stadler, editors. *Protocells: Bridging Nonliving and Living Matter*. MIT Press, Cambridge, MA, 2008.

[244] T. S. Ray. An approach to the synthesis of life. In C. Langton, C. Taylor, J. Farmer, and S. Rasmussen, editors, *Artificial Life II*, volume X of *SFI Studies in the Sciences of Complexity*, pages 371–408. Addison-Wesley, 1991.

[245] M. Redfield. *Phantom Formations: Aesthetic Ideology and the Bildungsroman*. Cornell University Press, Ithaca, NY, 1996.

[246] J. Reed, R. Toombs, and N. A. Barricelli. Simulation of biological evolution and machine learning: I. Selection of self-reproducing numeric patterns by data processing machines, effects of hereditary control, mutation type and crossing. *Journal of Theoretical Biology*, 17(3):319–342, 1967.

[247] J. A. Reggia, H.-H. Chou, and J. D. Lohn. Cellular automata models of self-replicating systems. In *Advances in Computers*, volume 47, pages 141–183. Elsevier, 1998.

[248] J. A. Reggia, J. D. Lohn, and H.-H. Chou. Self-replicating structures: Evolution, emergence, and computation. *Artificial Life*, 4(3):283–302, 1998.

[249] J. Rieffel, D. Knox, S. Smith, and B. Trimmer. Growing and evolving soft robots. *Artificial life*, 20(1):143–162, 2014.

[250] J. Riskin. *The Restless Clock: A History of the Centuries-long Argument Over what Makes Living Things Tick*. University of Chicago Press, Chicago, IL, 2016.

[251] S. Roberts.   From Homer to HAL: 3,000 years of AI narratives. *Research Horizons*, 35:28–29, Feb. 2018.   University of Cambridge. https://issuu.com/uni_cambridge/docs/issue_35_research_horizons.

[252] R. Robinson.   Butler, Samuel.   In *Dictionary of New Zealand Biography. Te Ara – The Encyclopedia of New Zealand.*   Updated 5 June 2013. http://www.TeAra.govt.nz/en/biographies/1b55/butler-samuel.

[253] S. A. Roe. *Matter, life, and generation: Eighteenth-century embryology and the Haller-Wolff debate.* Cambridge University Press, Cambridge, U.K., 1981.

[254] L. J. Rothschild, C. Maurer, I. G. Paulino Lima, D. Senesky, A. Wipat, and J. Head III.   Myco-architecture off planet: Growing surface structures at destination. Technical Report HQ-E-DAA-TN66707, NASA, Mar. 2019. https://ntrs.nasa.gov/search.jsp?R=20190002580.

[255] S. Russell, D. Dewey, and M. Tegmark.   Research priorities for robust and beneficial artificial intelligence. *AI Magazine*, 36(4):105–114, 2015.

[256] F. Saberhagen. *Berserker*. Ballantine Books, New York, 1967.

[257] C. Sagan. Direct contact among galactic civilizations by relativistic interstellar spaceflight. *Planetary And space Science*, 11(5):485–498, 1963.

[258] Science and Technology Committee.   Robotics and artificial intelligence.   HC 145. Published on 12 October 2016 by authority of the House of Commons, Oct. 2016.   https://publications.parliament. uk/pa/cm201617/cmselect/cmsctech/145/145.pdf.

[259] M. Shanahan. *The Technological Singularity*. MIT Press, 2015.

[260] R. Sheckley. The Necessary Thing. *Galaxy Science Fiction*, pages 55–66, June 1955.

[261] M. Shelley. *Frankenstein; or, The Modern Prometheus.* Lackington, Hughes, Harding, Mavor and Jones, London, 1818.

[262] C. Shulman and N. Bostrom. How hard is artificial intelligence? Evolutionary arguments and selection effects. *Journal of Consciousness Studies*, 19(7-8):103–130, 2012.

[263] M. Sipper. Fifty years of research on self-replication: An overview. *Artificial Life*, 4(3):237–257, 1998.

[264] M. Sipper and J. A. Reggia.   Go forth and replicate. *Scientific American*, pages 35–43, Aug. 2001.

[265] J. Sladek. *The Reproductive System*. Victor Gollancz, London, 1968.

[266] J. J. Smith, W. R. Johnson, A. M. Lark, L. S. Mead, M. J. Wiser, and R. T. Pennock. An Avida-ED digital evolution curriculum for undergraduate biology. *Evolution: Education and Outreach*, 9(1):9, 2016.

[267] L. J. Snyder. *The Philosophical Breakfast Club: Four Remarkable Friends Who Transformed Science and Changed the World.* Broadway Books, New York, NY, 2011.

[268] E. H. Spafford. Computer viruses as artificial life. *Artificial Life*, 1(3):249–265, 1994.

[269] K. O. Stanley, J. Lehman, and L. Soros. Open-endedness: The last grand challenge you've never heard of. *O'Reilly Ideas*, Dec. 2017. https://www.

oreilly.com/ideas/open-endedness-the-last-grand-challenge-youve-never-heard-of.

[270] O. Stapledon. *Star Maker*. Methuen & Co., London, 1937.

[271] O. Stapledon. *First and Last Men*. Millennium, London, SF Masterworks edition, 1999. (Originally published in 1930).

[272] O. Stapledon. *Star Maker*. Millennium, London, SF Masterworks edition, 1999. (Originally published in 1937).

[273] J. Suthakorn, A. B. Cushing, and G. S. Chirikjian. An autonomous self-replicating robotic system. In *Advanced Intelligent Mechatronics, 2003. AIM 2003. Proceedings. 2003 IEEE/ASME International Conference on*, volume 1, pages 137–142. IEEE, 2003.

[274] M. Szollosy. Freud, Frankenstein and our fear of robots: projection in our cultural perception of technology. *AI & Society*, 32(3):433–439, 2017.

[275] P. Szor. *The Art of Computer Virus Research and Defense*. Addison-Wesley, Boston, MA, 2005.

[276] T. Taylor. Creativity in evolution: Individuals, interactions and environments. In P. J. Bentley and D. W. Corne, editors, *Creative Evolutionary Systems*, chapter 1, pages 79–108. Morgan Kaufmann, 2001.

[277] T. Taylor. Redrawing the boundary between organism and environment. In J. Pollack, M. Bedau, P. Husbands, T. Ikehami, and R. Watson, editors, *Artificial Life IX: Proceedings of the Ninth International Conference on the Simulation and Synthesis of Living Systems*, pages 268–273, Cambridge, MA, 2004. MIT Press.

[278] T. Taylor. Evolution in virtual worlds. In M. Grimshaw, editor, *The Oxford Handbook of Virtuality*, chapter 32. Oxford University Press, 2013.

[279] T. Taylor. Evolutionary innovations and where to find them: Routes to open-ended evolution in natural and artificial systems. *Artificial Life*, 25(2):207–224, 2019.

[280] T. Taylor, J. E. Auerbach, J. Bongard, J. Clune, S. Hickinbotham, C. Ofria, M. Oka, S. Risi, K. O. Stanley, and J. Yosinski. WebAL comes of age: A review of the first 21 years of artificial life on the web. *Artificial Life*, 22(3):364–407, 2016. (doi:10.1162/ARTL_a_00211).

[281] T. Taylor, M. Bedau, A. Channon, D. Ackley, W. Banzhaf, G. Beslon, E. Dolson, T. Froese, S. Hickinbotham, T. Ikegami, et al. Open-ended evolution: Perspectives from the OEE workshop in York. *Artificial Life*, 22(3):408–423, 2016. (doi:10.1162/ARTL_a_00210).

[282] T. Taylor and A. Dorin. Past visions of artificial futures: One hundred and fifty years under the spectre of evolving machines. In *Proceedings of the Artificial Life Conference 2018*, pages 91–98. MIT Press, 2018.

[283] T. J. Taylor. *From Artificial Evolution to Artificial Life*. PhD thesis, University of Edinburgh, College of Science and Engineering, School of Informatics, 1999. (Available at https://www.era.lib.ed.ac.uk/handle/1842/361).

[284] M. Tegmark. *Life 3.0: Being Human in the Age of Artificial Intelligence*. Alfred A. Knopf, New York, 2017.

[285] The Staff, Electronic Computer Project. Report on Contract No. N-7-ONR-388, Task Order I. Technical report, The Institute for Advanced Study, Electronic Computer Project, Princeton, New Jersey, Apr. 1954.

[286] B. Thomas. Alfred Marshall on economic biology. *Review of Political Economy*, 3(1):1–14, 1991.

[287] K. Tingley. Evolution of machines. *Century Path*, 13(42):2, Aug. 1910.

[288] F. J. Tipler. Extraterrestrial intelligent beings do not exist. *Quarterly Journal of the Royal Astronomical Society*, 21:267–281, 1980.

[289] F. J. Tipler. Additional remarks on extraterrestrial intelligence. *Quarterly Journal of the Royal Astronomical Society*, 22:279–292, 1981.

[290] F. J. Tipler. A brief history of the extraterrestrial intelligence concept. *Quarterly Journal of the Royal Astronomical Society*, 22:133–145, 1981.

[291] F. J. Tipler. Extraterrestrial intelligent beings do not exist. *Physics Today*, 34:9,70–71, Apr. 1981.

[292] A. M. Turing. Intelligent machinery, a heretical theory. Unpublished typescript of lecture presented at a meeting of the "51 Society" at Manchester University in 1951. Republished in S. Barry Cooper and J. van Leeuwen (eds.) "Alan Turing: His Work and Impact", Elsevier, 2013. Page numbers given in the text refer to this republished version.

[293] A. M. Turing. On computable numbers, with an application to the Entscheidungsproblem. *Proceedings of the London Mathematical Society (Series 2)*, 42:230–265, 1937.

[294] A. M. Turing. Intelligent machinery. Technical report, National Physical Laboratory, 1948. Available at http://www.alanturing.net/intelligent_machinery. Republished in Copeland, J.B., editor (2004). *The Essential Turing*. Oxford University Press.

[295] A. M. Turing. Computing machinery and intelligence. *Mind*, 49:433–460, 1950.

[296] S. M. Ulam. John von Neumann 1903–1957. *Bulletin of the American Mathematical Society*, 64:1–49, 1958.

[297] S. M. Ulam. *Adventures of a Mathematician*. Scribner, New York, NY, 1976.

[298] V. A. Uspensky. Andrei Nikolaevich Kolmogorov — the great Russian scientist. In D. A. Pospelov and Y. I. Fet, editors, *Essays on the History of Computer Science in Russia*, pages 484–505. Scientific Publishing Center of the Siberian Division of RAS (the Russian Academy of Sciences), Novosibirsk, 1998. (In Russian).

[299] A. E. van Vogt. M 33 in Andromeda. *Astounding Science Fiction*, pages 129–142, Aug. 1943.

[300] P. A. Vargas, E. A. Di Paolo, I. Harvey, and P. Husbands. *The Horizons of Evolutionary Robotics*. MIT Press, 2014.

[301] N. Virgo, C. Fernando, B. Bigge, and P. Husbands. Evolvable physical self-replicators. *Artificial Life*, 18(2):129–142, 2012.

[302] J. von Neumann. The general and logical theory of automata. In L. A. Jeffress, editor, *Cerebral Mechanisms in Behavior. The Hixon Symposium*, pages

1–31. John Wiley & Sons, New York, NY, 1951. (The Hixon Symposium took place in September 1948).

[303] J. von Neumann. *The Theory of Self-Reproducing Automata.* University of Illinois Press, Urbana, Ill., 1966. Editor: A.W. Burks.

[304] A. Wagner. *The origins of evolutionary innovations: a theory of transformative change in living systems.* Oxford University Press, Oxford, UK, 2011.

[305] G. P. Wagner. Evolutionary innovations and novelties: Let us get down to business! *Zoologischer Anzeiger,* 256:75–81, 2015.

[306] P. S. Warrick. *The Cybernetic Imagination in Science Fiction.* MIT Press, Cambridge, MA, 1980.

[307] A. Weber and F. J. Varela. Life after Kant: Natural purposes and the autopoietic foundations of biological individuality. *Phenomenology and the Cognitive Sciences,* 1(2):97–125, 2002.

[308] H. G. Wells. *The War of the Worlds.* William Heinemann, London, 1898.

[309] L. L. Whyte. *Archimedes, or The Future of Physics.* Kegan Paul, Trench, Trübner & Co., London, 1927.

[310] L. L. Whyte. *The Next Development of Mankind.* Routledge, Abingdon, UK, 2017. (This edition with an introduction by Brian Rothery and Brian David. First edition published in 1944 by The Cresset Press).

[311] N. Wiener. *Cybernetics or Control and Communication in the Animal and the Machine.* MIT Press, Cambridge, MA, 2nd edition, 1961.

[312] R. M. Williams. Robots return. *Astounding Science Fiction,* pages 140–147, Sept. 1938.

[313] S. F. Wright. Automata. *Weird Tales,* pages 337–344, Sept. 1929.

[314] F. Zhang, J. Nangreave, Y. Liu, and H. Yan. Structural DNA nanotechnology: State of the art and future perspective. *Journal of the American Chemical Society,* 136(32):11198–11211, 2014.

[315] K. Zuse. Sich selbst reproduzierende Systeme. Konrad Zuse Internet Archive (http://zuse.zib.de), 1941. ZIA ID: 0423. Transcription (in German) of original handwritten stenographic notes.

[316] K. Zuse. Gedanken zur Automation und zum Problem der technischen Keimzelle. *Unternehmensforschung,* 1:160–165, 1957.

[317] K. Zuse. Über sich selbst reproduzierende Systeme. *Elektronische Rechenanlagen,* 9(2):57–64, 1967.

[318] K. Zuse. *The Computer — My Life.* Springer-Verlag, Berlin, 1993. English translation of the original German version "Der Computer — Mein Lebenswerk", 1993.

[319] V. Zykov, E. Mytilinaios, B. Adams, and H. Lipson. Robotics: Self-reproducing machines. *Nature,* 435(7039):163–164, 2005.

Printed in the United States
By Bookmasters